EL

OF

STEAM ENGINEERING.

BY

H. W. SPANGLER,

Whitney Professor of Dynamical Engineering in the University of Pennsylvania,

ARTHUR M. GREENE, JR.,

Professor of Mechanical Engineering in the University of Missouri,

AND

S. M. MARSHALL, B.S. IN E.E.

FIRST EDITION.

SECOND THOUSAND.

Elements of Steam Engineering

by H. W. Spangler, Arthur M. Greene and S. M. Marshall

ISBN 0-917914-14-7

1 2 3 4 5 6 7 8 9 0

PREFACE.

THIS book is intended to bring before the beginner examples of the various forms of steam apparatus used in modern steam-power plants, to explain simply and briefly the construction, use, and reasons for using these various parts or machines, and to give him a working vocabulary in this branch of engineering. The book is prepared, primarily, for first-year students in engineering schools, but it is hoped that it may be of use to the general reader and to the great number of young men in the manual-training schools and institutes throughout the country.

The work contains little theoretic matter, as, in most cases, these portions are taken in the latter part of a student's course, and after he has become familiar with many machines. In the early part of his course, however, he should be taught the uses of the various machines he finds about him.

In preparing the work, English and American works on the steam-engine and the technical papers have been freely consulted, though, in most cases, the drawings have been made from the actual machines, or from blue prints supplied by the builder. The authors are indebted to the manufacturers and engineers who have kindly furnished drawings and data and wish to express their appreciation and thanks.

THE AUTHORS.

January 6, 1903.

WARNING

Remember that the materials and methods described here are from another era. Workers were less safety conscious then, and some methods may be downright dangerous. Be careful! Use good solid judgement in your work, and think ahead. Lindsay Publications has not tested these methods and materials and does not endorse them. Our job is merely to pass along to you information from another era. Safety is your responsibility.

Write for a catalog or other unusual books available from:

Lindsay Publications
PO Box 12
Bradley, IL 60915-0012

CONTENTS.

ELEMENTS OF STEAM-ENGINEERING.

CHAPTER I.

BOILERS.

THE common method of utilizing, for commercial purposes, the heat in coal and other fuels is by the use of the steam-boiler and engine. In the boiler, water receives the heat of the fuel and is converted into steam. From the boiler, the steam is conveyed to the engine, where its energy is expended in driving a moving part so that some desired effect may be produced; as, for instance, turning the propeller-shaft of a steam-vessel, driving the wheels of a locomotive, the plunger of a pump, the armature of a dynamo, or the line-shaft of a manufacturing establishment. In any case, the heat energy is so transformed that it may be usefully applied.

Fuels.—The fuels in use for generating steam are coal, lignite, peat, wood, oil, coke, bagasse, and gas. **Coal** contains volatile hydrocarbons, which may be driven off by heat leaving **coke,** which is carbon in a porous state mixed with a small quantity of mineral matter. The carbon of the coke is termed the **fixed carbon** of the coal and the mineral matter is the **ash.** As the relative amounts of **volatile matter** and fixed carbon vary, so does the value of the coal, as is shown below.

Anthracite coal contains the lowest amount of volatile matter, there being from two to seven per cent in the anthracites of Pennsylvania.* This coal is hard, compact, and lustrous.

* The Calorific Power of Fuels, by Poole.

It burns with a short flame and, practically, no smoke. It usually contains more ash than other coals, and is apt to break into small pieces when heated. Some anthracites leave a hard, lumpy, mineral residue called **clinker,** while other varieties leave a soft ash. The ash in this coal varies from five to seventeen per cent, averaging about eleven per cent.

Bituminous coal contains from twenty-five to forty per cent of volatile hydrocarbons and from two to fourteen per cent ash, with an average of about six per cent. Some coals of this class have a very high ash content, but this is quite exceptional with American coals. It burns, usually, with a long flame and a strong tendency to smoke. Certain kinds soften on heating and flow into one large mass, which, if a strong fire is desired, must be broken up by means of a long, heavy bar termed a **slice-bar.** Such action in a coal is called **caking,** and the coals which exhibit it, caking bituminous coals, thus distinguishing them from the non-caking coals. The amount of volatile matter in bituminous coal gives us another classification—the **long-flaming** and the **short-flaming coal.** The former contains a larger percentage of volatile matter, and, as the air supply is insufficient to completely burn these gases at the surface of the coal, a long flame results. When this coal is first fired the volatile matter is driven off and, if heated air is properly supplied, the combustion takes place without much smoke. If, however, these gases are cooled before complete combustion occurs, a dense black smoke results. The residue, after the volatile matter is removed, is a porous coke, which burns very readily, giving an intense fire.

Between the bituminous and the anthracite coals occur two classes—the **semi-bituminous** and the **semi-anthracite** coal, which partake of some of the properties of the coals above mentioned.

Lignite, or **brown coal,** is high in volatile matter, moisture, and ash. It is coal in the process of formation and, though of lower heating value than anthracite or bituminous coal, is still of some worth in certain localities for fuel.

Peat.—In portions of Europe and Canada there are large deposits of vegetable débris called peat, which, although containing much water and some earthy matter, can, by treatment, be rendered available for fuel. It is of value in these localities because of the cost of other fuels. The deposits in the United States and Canada have not been worked to any extent because of the cost of preparation. The composition of peat after drying is practically the same as wood.

Wood.—Wood is used as fuel for steam generation in localities remote from the coal-fields, or where it is a waste product from a manufacturing process. It gives an intense fire and a small amount of light ash, with rather dense smoke. Its heating value is from forty to eighty per cent that of coal.

Oil.—Petroleum and petroleum waste are used extensively in Russia for fuel, and are being used to some extent in the western part of the United States. Oil is easy to apply and regulate, the firing is continuous and cleaner than with coal, and the heating value is high. It is, however, too costly for ordinary use in the eastern part of the United States because of the nature of the natural oil obtained. In Russia the oil is so different in character that its commercial use for fuel is possible.

Gas.—Natural gas has been used extensively as a fuel, but owing to the failure of the supply its use under boilers is diminishing.

Producer Gas.—In some localities poor coal is burned in an apparatus called a producer in such a way that the result of this incomplete combustion is a mixture of carbon monoxide, volatile hydrocarbons, hydrogen, and nitrogen. This gas, called producer-gas, has a lower heating power than natural or illuminating gas, but is of great value because poor fuels may be utilized.

Coke, Bagasse.—The by-products or refuse of many manufacturing processes are used as fuels. Coke from gas-retorts, crushed sugar-cane (bagasse) from the sugar-mill, sawdust from sawmills, and the waste gases from the blast-furnace are all used as fuels with success.

The following table gives the average heating value of fuels, the values being compiled from the works of Kent, Poole, and others:

HEATING VALUES OF FUEL.

Fuel.	British Thermal Units per Pound.	
Anthracite coal	14,600	
Semi-anthracite coal	14,700	
Semi-bituminous coal	15,500	
Bituminous coal	14,400	
Lignite	11,000	
Peat	9,000	
Wood (yellow pine)	9,153	
Wood (ash)	8,480	
Wood (oak)	8,316	
Oil	19,000	
Coke	14,300	
Bagasse	3,000	
Natural gas	1,000	per cu. ft. at atmospheric pressure.
Illuminating-gas	600	
Producer-gas	150	

Steam.—If heat be added to cold water its temperature rises, until, if the water be in a vessel open to the atmosphere, it reaches a temperature of 212 degrees Fahrenheit. After this temperature has been reached, the continued addition of heat causes no further rise in temperature, but steam will issue from the vessel. The water has been changed to a vapor occupying, at atmospheric pressure, about sixteen hundred times the volume of the water from which it was formed. This change is produced by the heat which has been added to the water, and which has been spent in separating the particles of water, transforming the liquid into a vapor. This vapor is similar to a gas, exerting on any vessel in which it is contained an expansive force. Should the water be placed in a closed vessel, it would begin to boil at 212 degrees Fahrenheit, but the steam first formed would be compressed as more is generated. This action raises the pressure in the vessel, and, at the same time, the temperature of the water and of the vapor rises, so that when enough water has been evaporated to make the pressure one hundred pounds per square inch, the temperature is 337 degrees Fahrenheit. If the attempt is made to heat water in a partial vacuum, it boils at a temperature much lower than 212 degrees Fahrenheit. It is a fact that a definite temperature corresponds to each pressure. The law relating to these has been determined experimentally, and tables constructed giving the corresponding values of pressure and temperature.

The heat produced by burning fuels is applied to the water contained in a vessel called a **boiler,** and, after raising the temperature of the water to that corresponding to the **boiling-point** for the existing pressure, this heat begins to change the water into steam. If this steam is formed as fast as it is allowed to leave the vessel the pressure within remains constant. Should a greater quantity be generated the pressure will rise, and with it the temperature of the water, while if too much steam is allowed to escape the pressure will fall. In the latter case a more intense fire must be maintained to evaporate a larger quantity of water and thus maintain the pressure. As the pressure on the steam becomes greater, the steam weighs more per cubic foot.

Boilers.—The earliest boilers used were of various shapes, but, as higher pressures were employed, it became necessary for structural reasons to use a cylindrical shape for the main portion or **shell.** If the ends of the cylinder, termed the **heads** of the boiler, were made as segments of spheres, the complete boiler could be made without any braces being required to maintain its shape, since, in this case, the internal pressure has no tendency to distort the shell or heads. This shell of iron containing water was supported on brick walls and a fire was built under one end, the heated products of combustion passing along and around the shell to the other end and thence to a chimney. Fig. 1 shows this form of boiler, (**plain cylindrical, externally fired**). The shell S, with the dished or spherical heads H, contains water to the level L.

This water is introduced through the **feed-pipe** F, and, on being made into steam, leaves by the **steam-nozzle** N. At V is the **safety-valve,** a valve which opens when the pressure rises above a certain point, allowing the excess of steam to escape. To remove the water from the boiler the **blow-off cock** B, located at the bottom, is opened. The fire is made on the **grate** G, and the products of combustion pass to the back of the boiler, then forward on one side, and, crossing over at the front, return on the other side to the back. The principal defect of this form of boiler was that it had a small amount of sur-

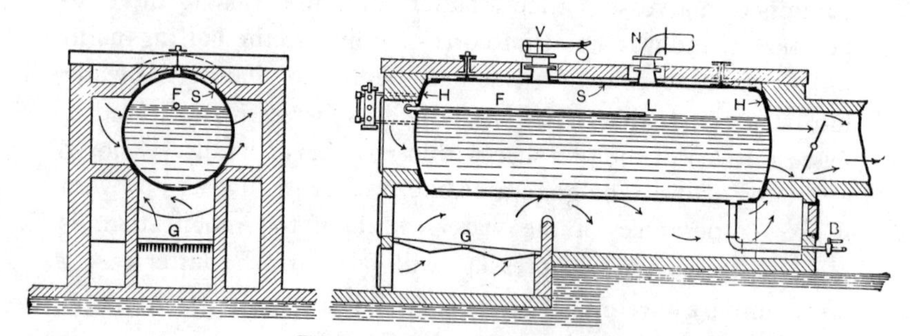

FIG. 1.—Plain Cylindrical Boiler.

face exposed to the heat, called **heating-surface,** and it required a large floor space for a given power. To remedy this, a large **flue** was introduced through the cylinder, and the grate, on which the fire was built, was placed within it. The gases then passed through the flue and thence around the boiler to the chimney. Later, two flues were introduced and, finally, the heating surface of these was increased by the introduction of cross tubes, known as **Galloway tubes.**

Galloway Boiler. **(internally fired, flue.)** — This boiler consists of a cylindrical shell made of steel, which has a flue A (Figs. 2, 4) passing from the back head to a sheet near the front of the boiler. To this sheet are attached two **furnaces** BB, which may be seen in Figs. 2, 4. These furnaces are attached to the front head of the boiler. The fire is built on the grate C in the furnaces. It receives the necessary air through the grate from the **ash-pit** D and through small holes in the furnace door E (Figs. 2, 3). As the coal burns, the gases pass over the **bridge wall** F, through the flue to the back of the boiler, thence around the shell, and past the **damper** G into the chimney-flue. A damper is used on all boilers to regulate the amount of coal burned, as, by closing it, the draught is diminished and the fuel consumption de-

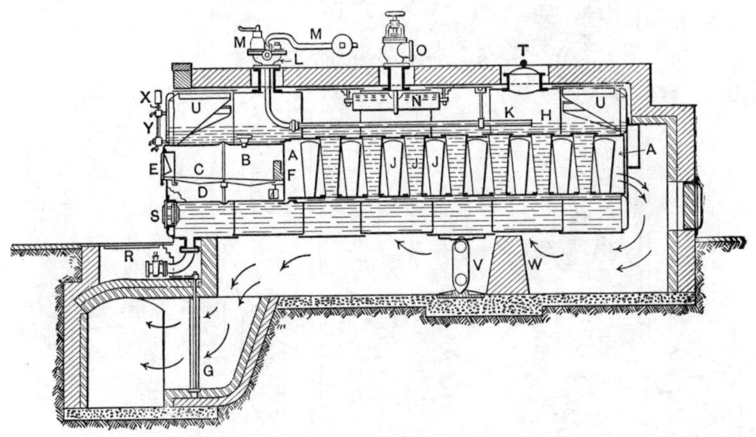

FIG. 2.—Galloway Boiler.

creased. The bridge F serves to mix the air and the combustible gases, thus making combustion more perfect, as well as forming a back for the fire-box.

The water is carried above the top of the flues to the level H, surrounding the various heated portions and entering the Galloway tubes J. It is introduced through a feed-pipe K which enters the boiler through a casting L to which the safety valves MM are attached. The steam formed is collected in the large **dry pipe** N, which is closed at the ends and contains a number of slots in its upper surface. The steam, not having a straight path from the surface of the water, and, moreover, having a large distributed area through which to enter, will have such a low velocity that there will be little water carried into the **steam line.** From the dry pipe the steam passes through the **stop-valve** O to the steam line. To remove the scum which may form on the surface of the water, a pipe, called a **surface blow,** similar to the feed, is attached on the other side of the safety-valve outlet. This is controlled by the valve P, Fig. 3, the feed being controlled by Q. When

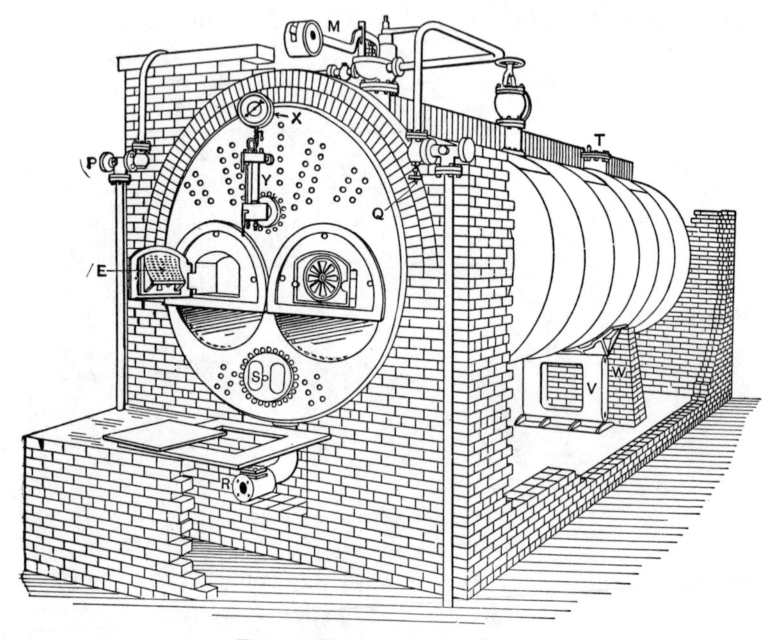

FIG. 3.—Galloway Boiler Setting.

it is desired to remove the scum from the boiler, P is opened and the steam pressure forces the scum out through the pipe. To clean the boiler entirely, another blow-off called the **bottom blow** is placed at the bottom at R. After removing the water, the boiler may be entered by either of the **manholes** S or T.

To prevent the main flue from collapsing, to increase the heating surface, and to aid circulation the Galloway tubes are introduced. These are conical tubes, formed as shown in

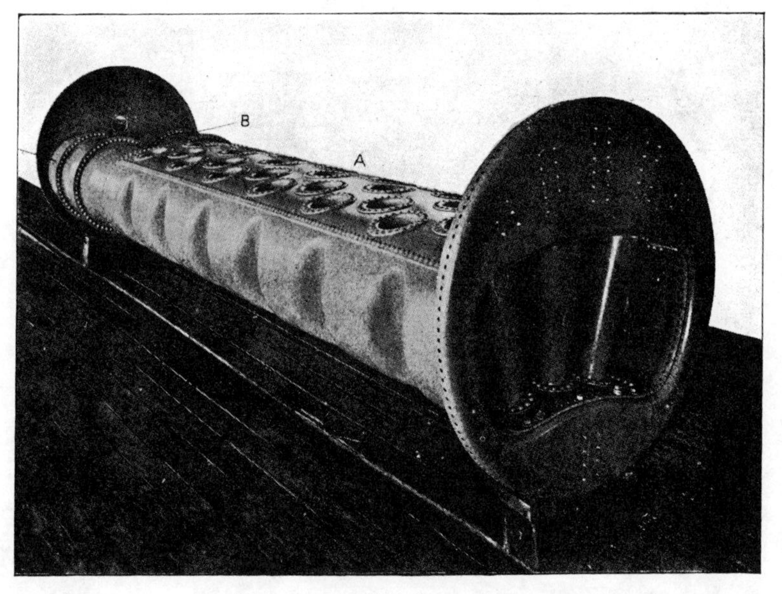

FIG. 4.—Flue of Galloway Boiler.

(*To face page* 8.)

Figs. 4, 5, with flanges at each end. They are introduced from the upper side of the flue and riveted before the flue is put in place. The side of the flue is stiffened by corrugations, as shown in Fig. 4. This figure shows the arrangement of Galloway tubes, which are so placed that the gases have not a straight path, the corrugations serving the same purpose. The flues are said to be **staggered** when arranged as shown in this figure. The furnaces are stayed by being flanged at short intervals, the one in Fig. 2 being shown with flanges and ring at the center of its length. To prevent the head sheets from bulging out from the steam pressure they are stiffened by means of **gusset stays U.** These consist of **plates** riveted to **angle irons** which are attached to the shell and head.

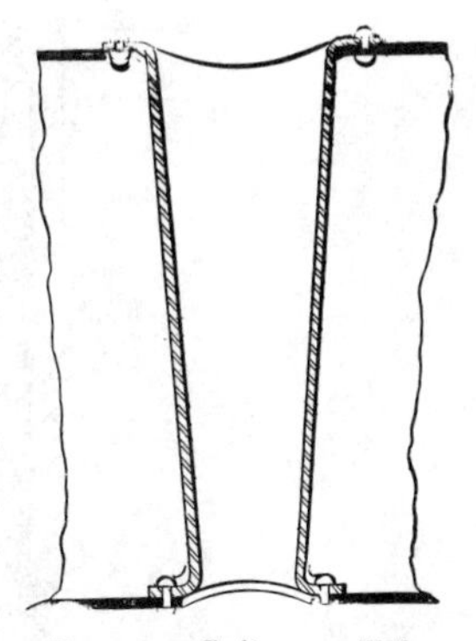

FIG. 5.—Galloway Tube.

The circulation of water in the Galloway boiler is upward from the furnaces and through the Galloway tubes, and downward at the sides. The hot gases pass underneath and, by heating the water at the bottom, aid in the circulation. The boiler is supported at the front by brick masonry, the back being carried on an **expansion rocker** V which allows the boiler to contract and expand. The brick setting is entirely free except at the front, so that there is no danger of injuring the setting as the boiler changes length. Behind the expansion rocker is built a safety pier W which is so near the shell that, in case the rocker breaks, the weight of the boiler will be carried by the pier without damage to the shell.

On the front of this boiler are the **steam-** and **water-gauges** X and Y, which show the steam pressure and the water-level.

Return Tubular Boiler **(externally fired, tubular).** —Another method of increasing the heating-surface of a plain cylindrical boiler is by adding a number of tubes through which

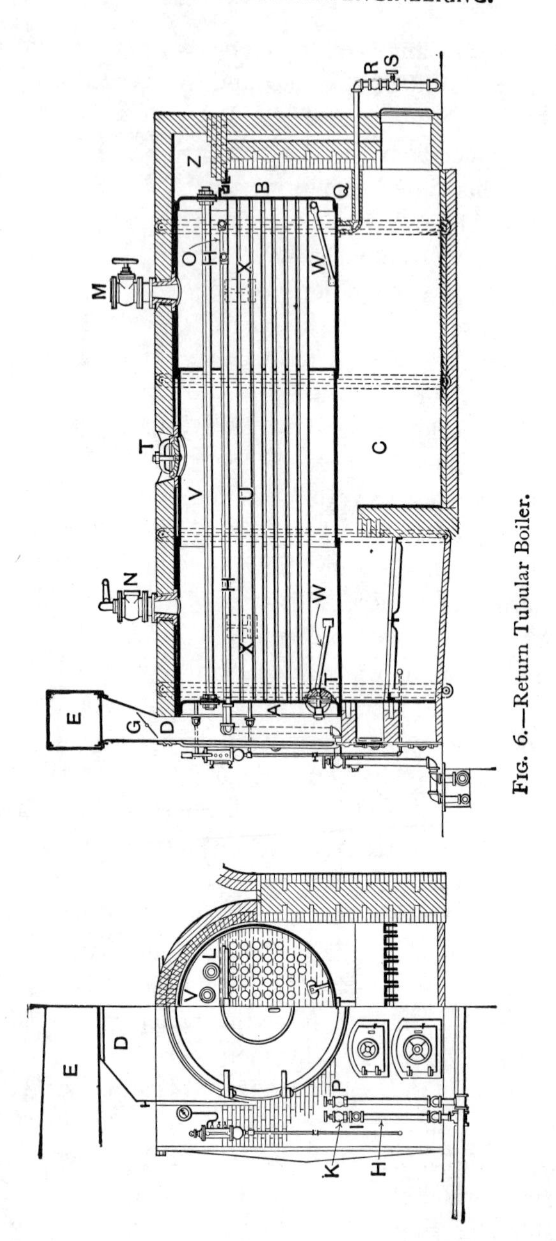

Fig. 6.—Return Tubular Boiler.

the gases, after passing along the bottom of the shell to the back, return to the front. This gives the return tubular boiler shown in Fig. 6. It is a **fire-tube** or tubular boiler in contradistinction to the **water-tube** or **tubulous** boiler. It consists of a shell containing a number of tubes extending from the **front tube-sheet** A to the **back tube-sheet** B. The products of combustion leaving the furnace above the grate pass over the bridge wall into the **combustion chamber** C, and thence through the tubes to the **uptake** D, entering the flue E. The damper G in the uptake serves to regulate the amount of fuel consumed.

Water enters through the feed-pipe H, the **check-valve** I, and the **regulating-valve** K. A check-valve, which allows a passage in one direction only, should be placed in the feed line where it enters the boiler, to prevent a discharge should the pressure in the feed line fall below that in the boiler. The water, which is carried to the level L, is distributed by the pipe H at the back of the boiler and, falling to the bottom, aids the circulation, which is upward at the front and downward at the back. The steam is taken off at the stop-valve M, near the back of the boiler. The safety-valve is located at N. The surface blow O is controlled by the valve P at the front of the boiler and at the back of the boiler is the bottom blow Q. This is located so that it will drain the boiler and is incased in firebrick to prevent the pipe burning out. The blow-off pipe Q usually contains a **cock** R, and sometimes a special valve S is added. The valve S is a special valve of such a form that there is little danger of scale collecting under the valve disc.

For cleaning, the boiler may be entered at either of the manholes TT, and the scale and other deposits removed from the shell or tubes. Most water contains mineral impurities and these may be precipitated from the water on boiling, so that, if not removed, they will form deposits in the boiler which may lead to disastrous results. Some of these deposits are soft and may be removed by blowing off an amount of water each day, while others form hard coatings, called **scale,** which can only be removed by scaling tools.

The steam pressure, which tends to force out the heads of the boiler, is resisted by the tubes U, the **through stays** V, and the **diagonal stays** W.

The boiler is supported by brackets X resting on the brickwork, the back brackets being free to move as the boiler expands. The brickwork is entirely free from the boiler, so that there is no danger of cracks being formed from the expansion of the shell. To allow for expansion, a joint is made with the setting at the back of the boiler by an iron Z bar dipping into a trough of sand, as shown at Z.

The furnace and ash-pit doors, as well as the large door used for cleaning the tubes (**front connection,** or uptake door), are shown in Fig. 6. This figure shows the water-column and steam-gauge. The tubes in this boiler are generally arranged in vertical rows and often with a wider space between the central rows. The setting of the boiler is made of brick masonry and is lined with **fire-brick** where it is subject to the direct action of the hot gases. To prevent loss of heat by radiation, the walls are often made double with an air space between, as shown in the back wall, Fig. 6.

Vertical Boiler (**internally fired, tubular**).—A type of boiler often employed for temporary work, or in places where floor space is limited, or where a self-contained boiler is required, is that shown in Fig. 7. It is formed of a cylindrical shell containing a **fire-box** A at the bottom. The top of the fire-box (**crown-sheet**) is connected by tubes to the upper head B. The fire-box is formed by a circular sheet Q riveted at one end to the crown-sheet C, often called the bottom tube sheet, and at the bottom to the **mud-ring** P. The space between the sheet Q and the shell is termed the **water-leg** D. This is necessary, as the metal around the fire must be in contact with water to prevent it from burning out. The fire-box side sheet and shell are stayed together by the **stay-bolts** EE, and an opening for firing is formed by a wrought-iron

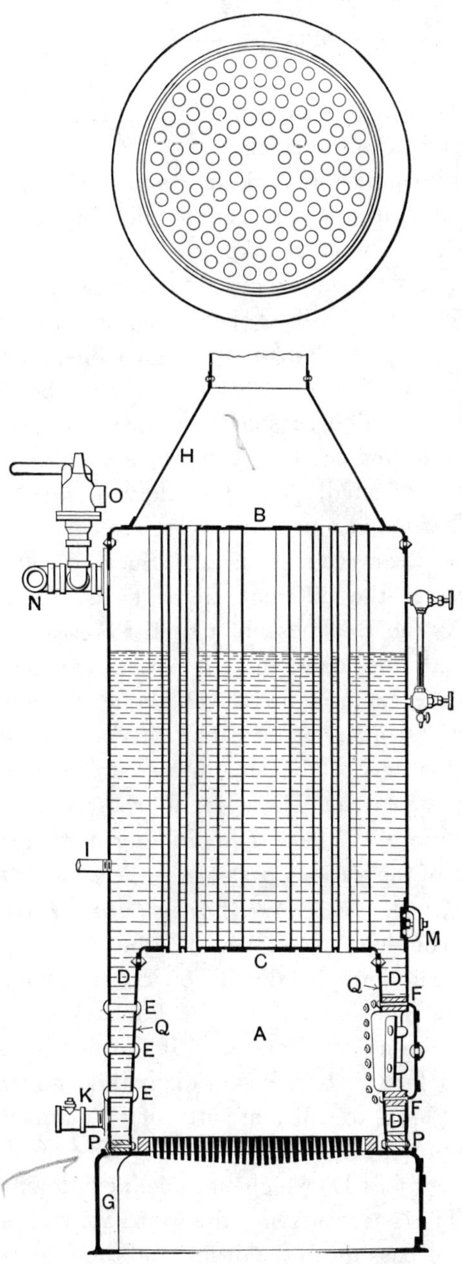

FIG. 7.—Vertical Boiler.

ring FF, as shown. The ash-pit is formed by a casting G, on which the shell and the grate rest. The **stack** rests on the **smoke-box** H, which is carried by the shell, the gases passing through the tubes to the stack.

I is the feed-pipe and K the bottom-blow. Hand-holes M are located at convenient places for cleaning the crown-sheet and water-leg. Steam is taken from N. O is the safety-valve. The only staying required for this boiler, except that of the side sheet of the fire-box, is accomplished by the tubes.

Water-tube, or ***Tubulous Boilers.***—Of late years there has been a tendency to use the water-tube boiler instead of the fire-tube. The reasons for this have been the use of higher pressures and the desire for the subdivision of the boiler into a number of small parts for safety. In this newer form the water is carried in tubes, and these are connected to a drum in which the steam can separate. Such a boiler is shipped conveniently, as the different elements are small and easily handled. As the thickness of a shell to resist a given pressure varies with the diameter, the parts can be made light, as all the diameters are small. The boiler is a quick steamer, since it contains little water; but this also renders it more liable to fluctuations of the steam pressure than a boiler with large water capacity. There is no reason for one type of boiler being more efficient than another if there is a proper arrangement of the various parts.

Babcock & Wilcox Water-tube Boiler (**externally fired, tubulous**).—Of the numerous forms of water-tube boilers, differing only in details of construction, but giving practically the same efficiencies, the Babcock & Wilcox Boiler (Fig. 8) is taken as an example. It consists of a **drum,** A, from thirty to forty-two inches in diameter, and from eighteen to twenty feet long, to which are attached a number of **headers,** B, by means of the connecting-pipes C. The headers are connected by the tubes D, which are inclined downward toward the back. The gases leaving the grate are compelled by the front **tiles** E to pass up to the drum, and then, by the hanging-

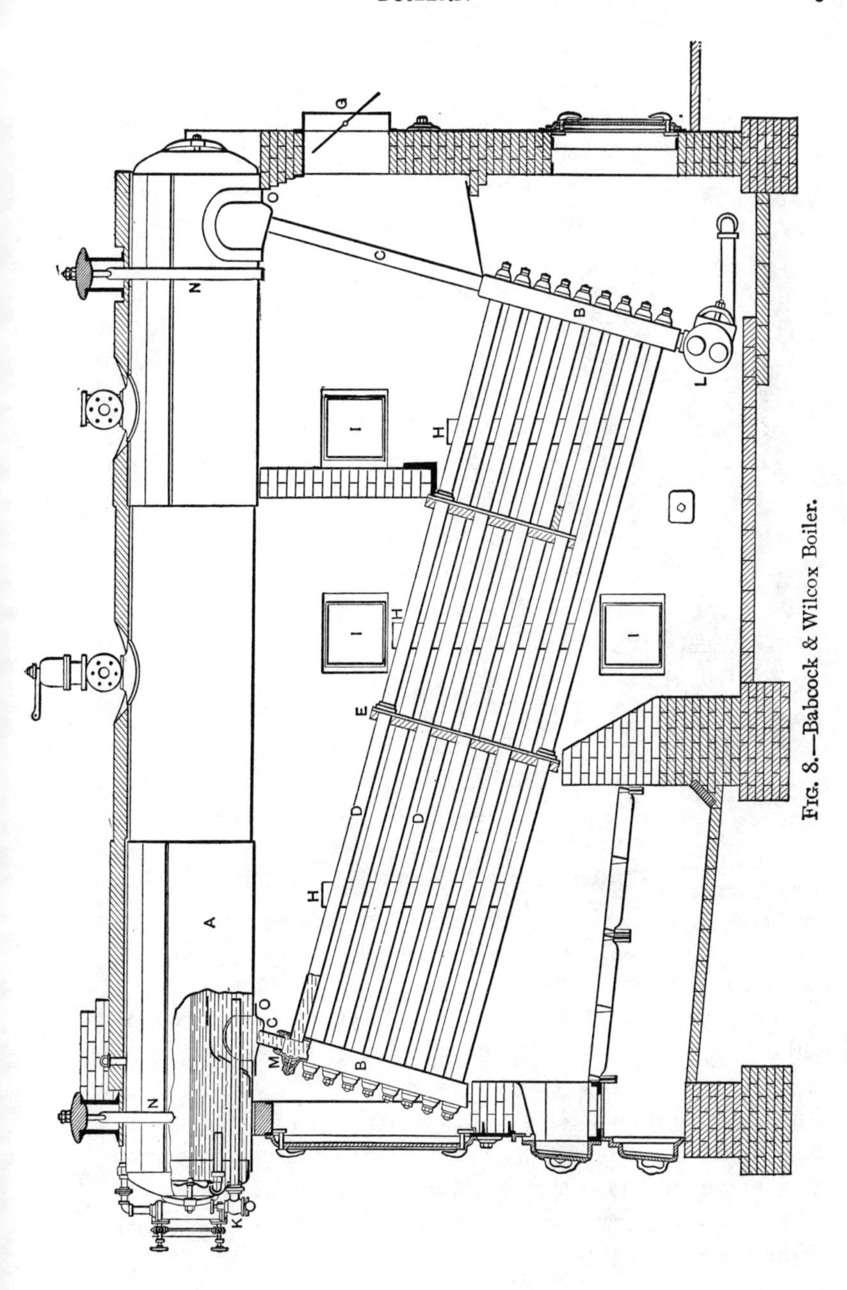

Fig. 8.—Babcock & Wilcox Boiler.

wall and back tiles, across the tubes again to the bottom. From this point they again cross the tubes and pass the damper to the flue G. To remove from the tubes the soot and ashes which may be carried over, **cleaning doors** HH are provided. Through these a steam-jet is introduced by means of a hose, thus reaching all parts of the tubes. The doors I are for the inspection of the drum A and tubes D.

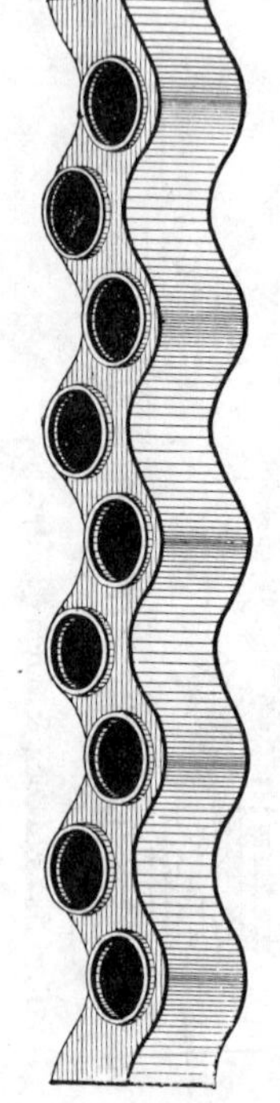

FIG. 9.—Babcock & Wilcox Header.

The water enters at K, as shown in the figure; it then flows down the back pipes and through the tubes to the front. Steam is taken from the back of the drum. At the back of the boiler is the **mud-drum** L, in which the sediment is partially collected, and from which it is removed by the bottom-blow.

The headers for this boiler are shown in Figs. 8 and 9. They are cast-iron or forged steel boxes, which have holes on one side for the tubes, and opposite these are holes fitted with **caps** M, which are held tight by T-headed bolts. The joint between the cap and the box is made tight by the surfaces being ground together or by the introduction of an inside hand-hole plate and an asbestos gasket. On the tops of these boxes are holes into which short pieces of tube called **nipples** are expanded, thus joining the headers to a saddle-box O, which is a single piece of pressed steel.

To clean the boiler the caps are removed from each end and a scraper run through the tubes. By means of the man-hole the drum is inspected or cleaned. The heads of the drum being dished steel plates, this boiler requires no staying. The drum is carried from a cross-girder by means of the straps N. These girders are carried on iron columns, and, as the tubes are swung from the drum, the whole boiler is independent of the brickwork.

Gill Water-tube Boiler **(water-tube, tubulous).**—Another of the water-tube boilers similar in general design to the one just described is the Gill water-tube boiler. It differs from it in having a series of small header-boxes (A, Fig. 10) instead of having a long, narrow header. These boxes are made with openings in one face for four, five, or six tubes, and opposite these are cleaning-holes, which are covered by the caps B, as shown.

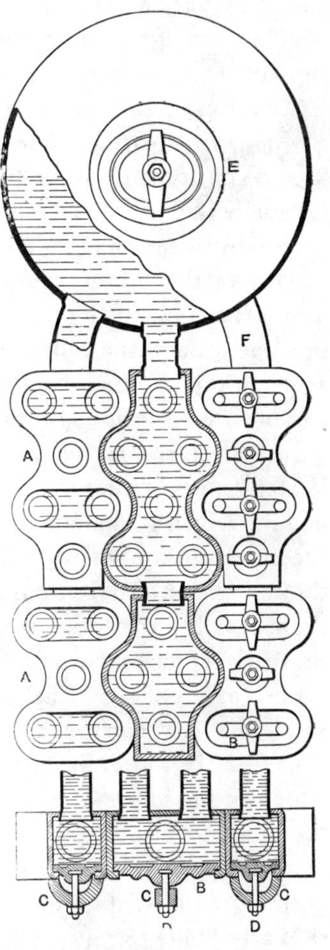

FIG. 10.—Gill Header.

The joint between the header-box and cap is an internal one. The cap is larger than the opening in the box, and, when entered by turning it on its face, it fits over the inside edge of the opening and is pressed out by the steam pressure. To make this joint steam-tight a piece of canvas, indurated with rubber, is placed between the two surfaces. This is called a **gasket,** which is the name given to any yielding material placed between two surfaces to make a steam-tight joint.

The yokes, or dogs C, together with T-headed bolts D, which fit in grooves in the cap, serve to keep this joint tight when the steam is off of the boiler.

When a cap is fitted on the outside of the header the steam pressure tends to relieve the pressure between the surfaces in contact, while with the joint on the inside, as used in this boiler, the steam pressure tends to make the joint tighter.

The boxes A are united by short nipples, which, with the other pipes, are made tight by being expanded into the holes in the box. The header-boxes are finally united to the drum E by the pipes F. The power of any boiler of this type is varied by varying the kind and number of these headers.

The arrangement of the drums and tubes, the method of support, the manner of placing the grate and passes for the gases are the same as those shown in Fig. 8.

Cahall Boiler **(externally fired, tubulous, vertical).** —A recent type of water-tube boiler with vertical tubes is the Cahall Boiler (Fig. 11). It consists of two drums, A and B, united by a series of tubes C, which diverge from the upper surface of the lower drum. The upper drum is made in the form of a ring, so that the gases, after passing from the fire-box and around the tubes, leave the boiler through the central space in this drum. The **baffle-plates** DD prevent them from passing directly through the open space at the center. The gases never reach the under surface of the lower drum A. Water enters A through the feed-pipe E, depositing the mud or salts which it contains. Rising through the tubes, it enters the upper drum, where steam is liberated, and then descends through the **external circulating pipes** (or **downtake**) F to the drum A. By this arrangement good circulation is produced. Steam is taken off at G. The blow-off H is attached to the lowest point of A. Each drum is provided with a manhole, so that access can be had for cleaning, and small holes are placed in the top of drum B for the removal of defective tubes. To clean the tubes a scraper is hauled back and forth through the tubes by means of a chain.

The lower drum has its head so dished that, except for the very large sizes, it requires no staying. The upper drum is similarly built, except that on the inner sheet a stiffening ring of angle-irons is fastened.

The boiler is supported on four cast-iron brackets attached to the lower drum. These rest on the masonry foundation, and are not shown in the figure.

Locomotive Boiler **(internally fired, tubular).**— The locomotive boiler has not been changed materially for a

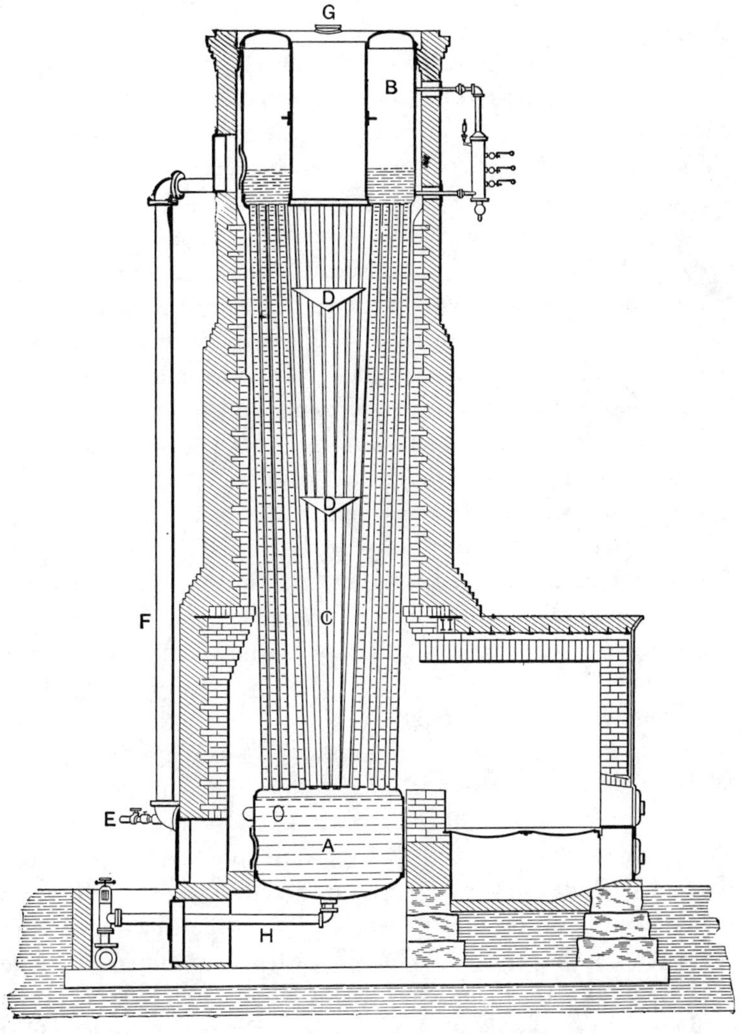

FIG. 11.—Cahall Boiler.

number of years, the improvements being mainly in details rather than general design. The requirements are a small boiler with large heating-surface, and a grate on which the fuel can be burned at a very high rate. To meet these requirements a great number of small tubes placed near together is used, and a strong draft is produced by the exhaust steam.

The locomotive boiler (Fig. 12) consists of a shell, A, com posed of several rings of steel, to which, at one end, is connected the fire-box B, and at the other the smoke-box C. From the front tube-sheet D, to the back tube-sheet E, are the tubes F arranged, as shown in Fig. 13, to increase the number of tubes, and thus the heating-surface. The fire-box is made up of two side-sheets, a back-sheet, and a tube-sheet, riveted together and to the crown-sheet over the top. Whenever possible, however, the side-sheets and crown-sheet are made of one piece of metal. These sheets are surrounded by water and form the best heating-surface of the boiler. Outside of these are two outside sheets, a back-sheet and a **throat-sheet** G, which, with the top-sheet, are riveted to the shell. The fire-box sheets and the outside sheets are riveted at their lower edge by long rivets to a forging, H, called the **water-space frame.** This closes the space between the sheets, forming the water-leg. On account of the arrangement of the outside of the shell over the fire-box this type is known as a Wagon-top Locomotive Boiler.

The fuel is fed through the door I upon the fire on the grate K, and the gases, rising from the fire, pass through the tubes to the smoke-box. The grate shown in Fig. 12 is a shaking-grate, used with soft coal. The separate grates are shaken by means of the rods J, which are attached to projections on each. Water enters at L through a check-valve, and steam is taken through the throttle-valve M and the **dry-pipes** N to the front of the boiler. Though called a dry-pipe, the pipe serves simply as a steam conduit and not as a steam separator, as does the dry-pipe shown in Fig. 2. From the front end of the dry-pipe the steam is distributed to the two cylinders by the two bent **steam-pipes** O. There is usually a blow-off cock at a low point in the water-leg. On the level with the crown-

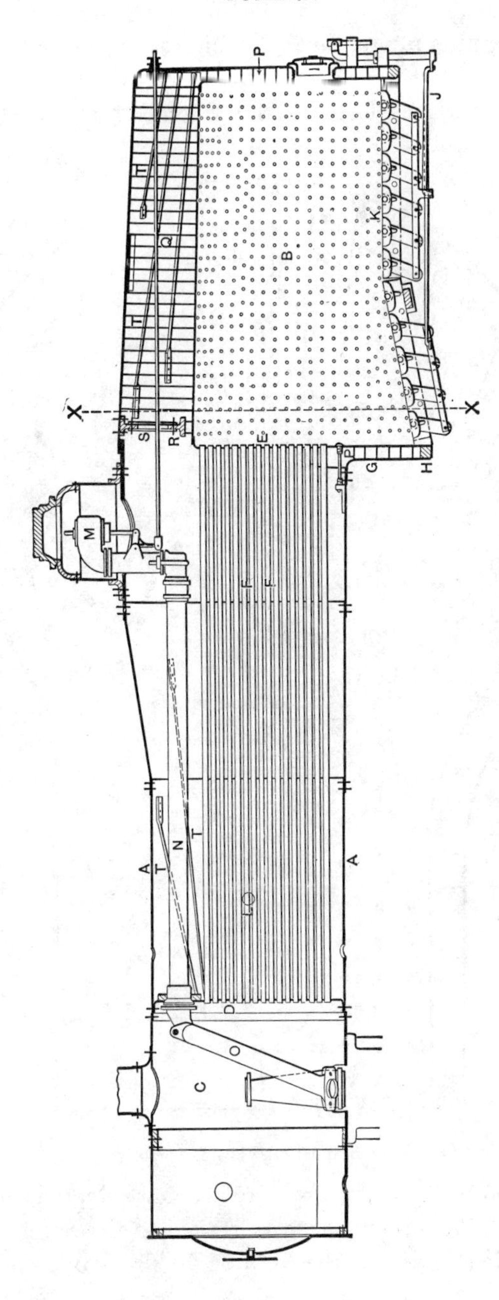

FIG. 12.—Locomotive Boiler, Wagon-top.

sheet are **cleaning plugs,** through which scale and sediment may be removed.

The flat surfaces forming the water-leg are supported by

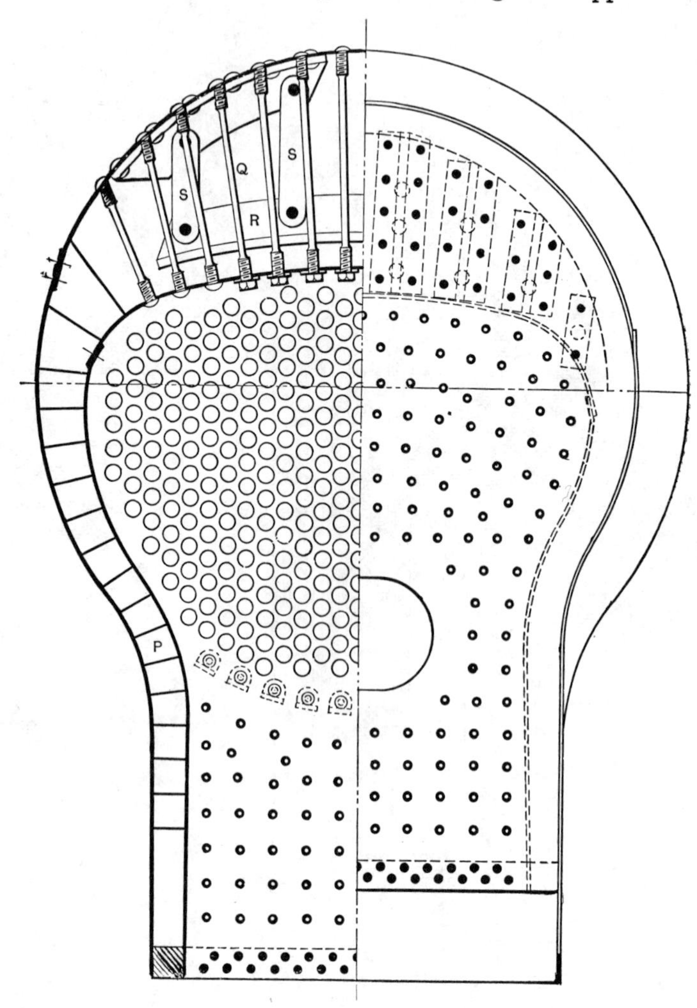

FIG. 13.—Section and End View Locomotive Boiler.

short stay-bolts P, which are screwed and riveted to each of the sheets. The crown-sheet is stayed to the outer sheet by the **radial stays** Q. At the front of the fire-box is shown another method of staying the crown-sheet by means of a **crown-bar** R and **sling stays** S. This method is sometimes used over the whole sheet, but in the boiler shown it is employed at the front only to allow for the expansion of the tube-sheet. The heads are supported by means of diagonal stays T.

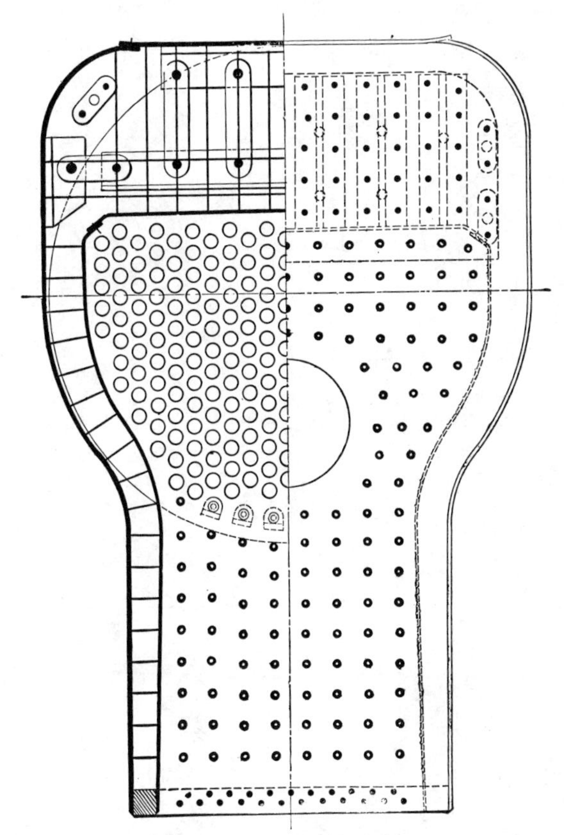

FIG. 14.—Belpaire Fire-box.

The locomotive boiler is bolted at the front to the saddle cast on the steam-cylinders, and at the back is supported by expansion pads riveted to the fire-box and resting on the side-frames, thus allowing for expansion. Fig. 13 shows a section through the fire-box on XX, Fig. 12, and a view of the fire-box end of the boiler. The arrangement of the radial stays Q, the sling stays S, and the stay-bolts is shown.

Special Fire-boxes.—In order to make the staying more simple, the **Belpaire fire-box** was introduced. In this form of fire-box, as shown in Fig. 14, the sheets are parallel and the stays are perpendicular to each surface, thus giving direct tension. In the wagon-top boiler the stays are normal

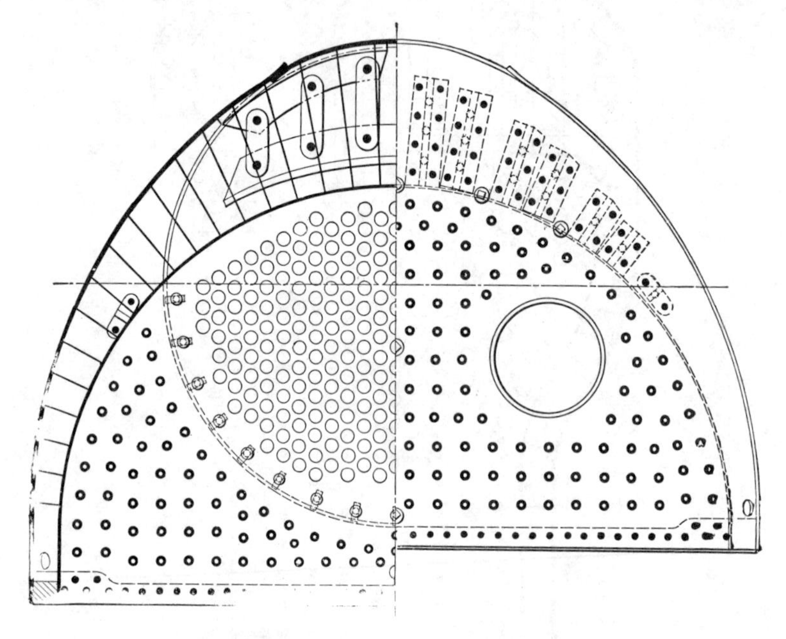

FIG. 15.—Wooten Fire-box.

to neither of the surfaces, and hence complex stresses are introduced.

To burn culm, the fine waste coal which results from the mining of anthracite coal, the **Wooten fire-box** was invented. This fire-box is so arranged that the grate may be made two or more times as large as the ordinary grate. With this grate the draft is gentle and sufficient to burn the proper amount of coal. With such fine coal the **rate of combustion** (coal per hour per square foot of grate-surface) must be low, as with a

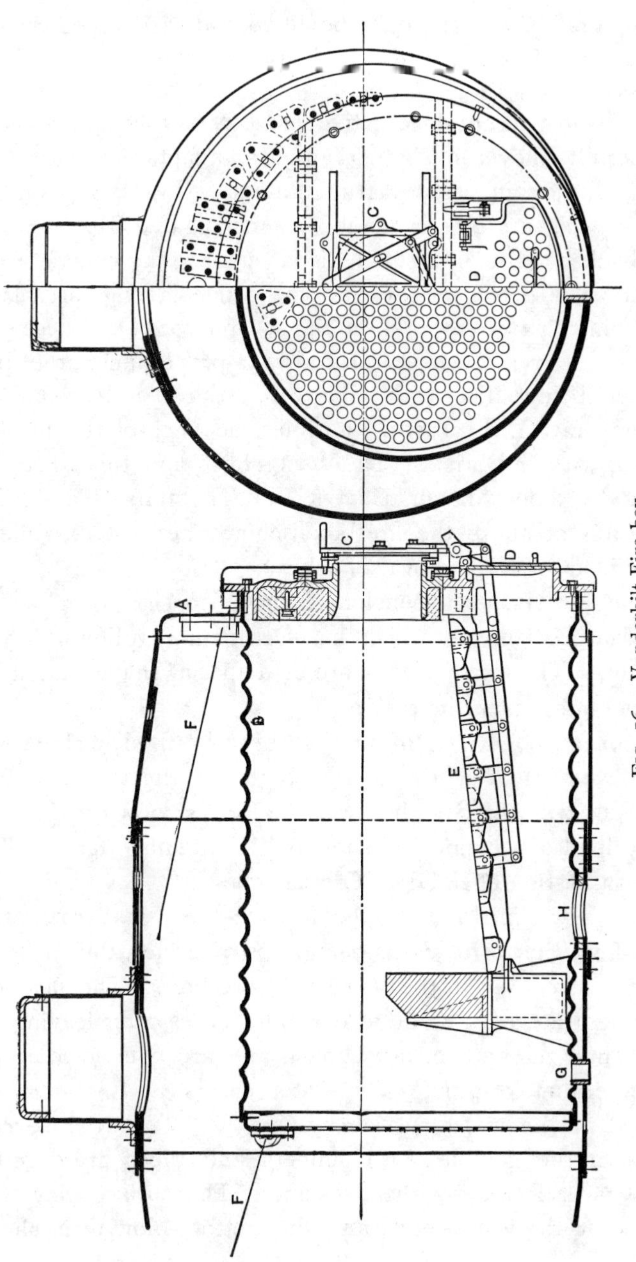

FIG. 16.—Vanderbilt Fire-box.

strong draft the coal would be lifted and discharged through the stack. Fig. 15 shows the construction of this type of fire-box.

The most recent locomotive fire-box is that known as the Vanderbilt, shown in Fig. 16, and is an adaptation of the common form now in use in marine boilers. No staying is necessary, as this fire-box is made of a corrugated Morrison flue. The back-head A of this boiler is fastened to the outside sheets and the flue B. Over the end of the flue castings are placed containing the fire-door C and the ash-pit door D. These castings are covered by sheet iron to give a proper finish, the space between being filled with asbestos. Above and around the rocking-grate E the casting forming the back of the fire-box is lined with fire-brick. The back-sheet and tube-sheet are stayed by a few diagonal stays F. The tube G is for the admission of air to the fire-box behind the bridge, while H is provided for the removal of ashes.

The **Morrison suspension furnace** consists of a welded flue of large diameter, which has corrugations rolled in it after welding. The corrugations are of a special shape and stiffen the flue and prevent its collapse.

Scotch Marine Boiler **(internally fired, tubular).**— The development of the marine boiler of the fire-tube type has resulted in the Scotch marine boiler, as shown in Fig. 17. It is formed of a cylindrical shell AA, containing furnace flues FF, combustion chambers CC, and tubes T.

As shown in Fig. 17, the furnace-flue F is a **Fox corrugated flue,** the corrugations being arcs of circles, thus differing from the flue used in the Vanderbilt fire-box. The flue contains the grate B and bridge D. The gases after leaving the tubes enter the uptake, not shown, and are conducted to the **funnel** or **smoke-pipe**. The water enters at E and is distributed at the water surface by pipe E. The water-level is such that the combustion chambers and tubes are covered. Steam is collected by the dry-pipe. The surface-blow J is located at the water-level, and the bottom-blow G is shown

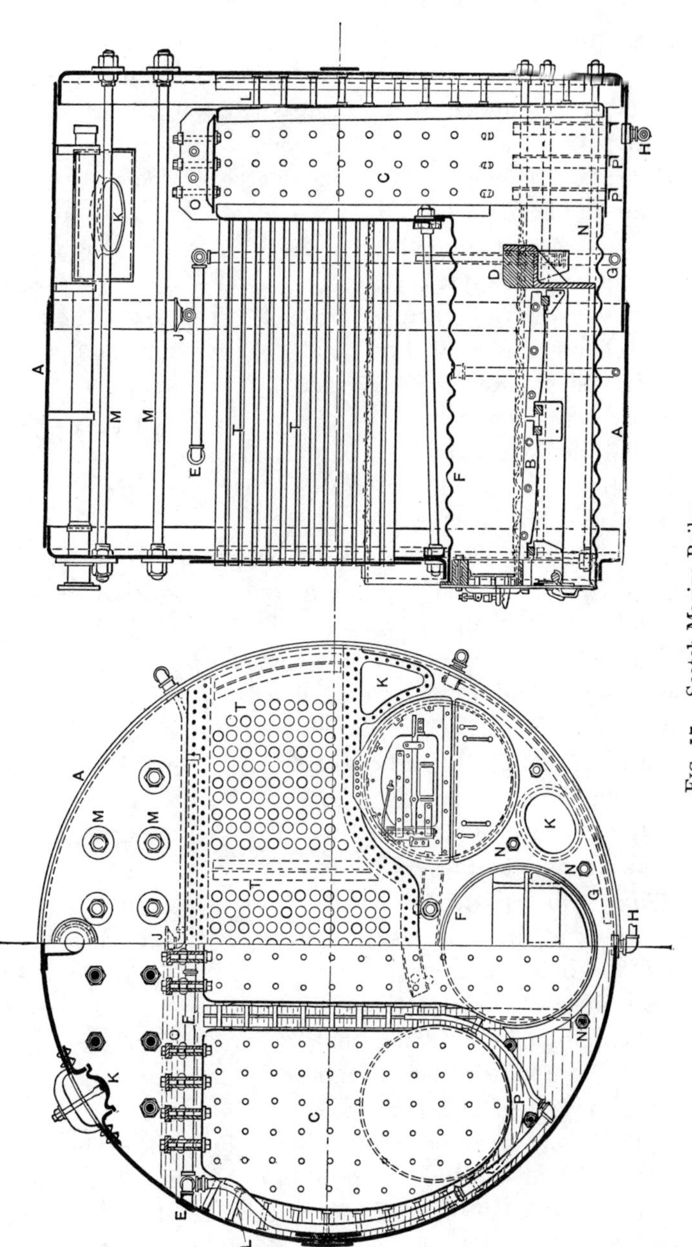

FIG. 17.—Scotch Marine Boiler.

leading to the bottom of the shell. To drain the water from the boiler when there is no steam pressure, a drain-pipe H is attached to the lowest part of the shell. Manholes K are provided so that all parts of the boiler are accessible for inspection and cleaning. The flat surfaces of the combustion chambers and heads are stayed by means of stay-bolts L and the tubes T. Above the combustion chamber long through stays M hold the opposite heads together. Such stays are also used below the combustion chamber at NN. The top of the combustion chamber is stayed by means of stay-bolts and girder-stays O, which latter are supported by the side-sheets. The lower portion of the combustion chamber is stiffened by means of angle-irons P riveted to the bottom sheet.

The boiler is carried in a saddle, which is riveted to the frames of the ship, as shown in Fig. 96.

The tubes are arranged in vertical rows and are connected to the combustion chamber in the rear. In some boilers there is a combustion chamber common to two flues, and where three flues are used only one combustion chamber may be employed. In the boiler shown in Fig. 17 each flue has a separate chamber.

A double-ended Scotch marine boiler is one in which furnaces are placed at each end of the boiler, generally with combustion chambers for each end separated by a water-leg. It is equivalent to two single-ended boilers placed end to end with the back heads removed.

Thornycroft Boiler **(water-tube)**.—As higher pressures were used, the thickness of metal required for the shell of the Scotch boiler became very great, and this, in addition to the necessity for reducing the weight of marine boilers, resulted in the employment of water-tube boilers. There are many forms of these in use to-day, and others have been proposed. As an example of this class, and to show the use of bent tubes, the Thornycroft boiler is shown in Fig. 18. This boiler has three drums—a large central drum A and two smaller lower drums BB. A number of bent tubes C of small diameter extend from A to B. These are so arranged that the outer

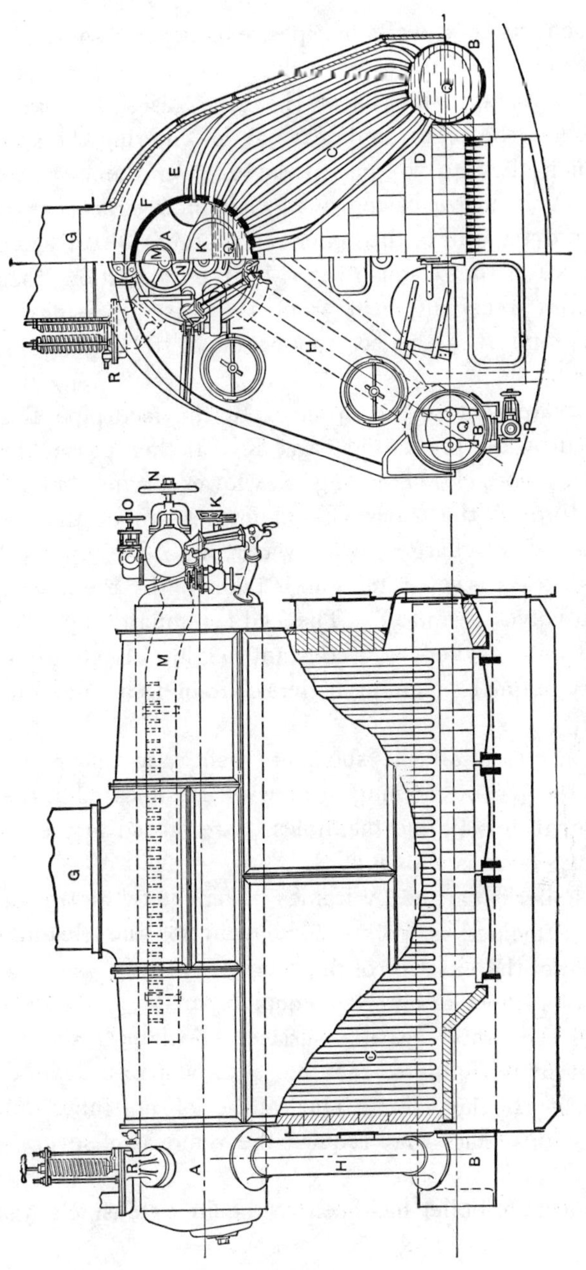

FIG. 18.—Thornycroft Boiler.

rows touch, forming walls of pipes, except for a short distance at the lower end of the furnace side, as at D, and at the upper end of the uptake side, as at E. The gases enter at D and pass around the tubes, and leave at E, entering the space F, from which they go to the uptake G, over the center of the boiler. A sheet-iron casing lined with some refractory material forms the outer wall of the boiler. The tubes directly over the fire receive a large quantity of heat by radiation, those on the interior receiving heat from the gases passing. The drums A and B are also connected by tubes H of large diameter.

The water enters the drum A by the feed-pipe I, and is controlled by the regulating valve K. It then passes through the outer tubes H, reaching the lower drum. The water ascends through the tubes C, and in these tubes the steam is formed, and discharges, with water, into the upper drum. Here the steam is separated and is taken off by the dry-pipe M and stop-valves N and O. The feed is regulated by a float, so that the water is kept at a constant height. Such a device is necessary in quick-steaming boilers containing a small quantity of water.

The heads of all the shells are dished, so that no staying is necessary. For cleaning, a blow-off P is located at the bottom of the drums B, and manholes Q are placed in each drum. The safety-valve is shown at R.

The boiler is carried by frames under the two lower drums. Doors are located at points convenient for the cleaning and inspection of the exterior of the tubes.

The bent tubes allow for unequal expansion and give an increase in the water-heating surface. Although these tubes have certain advantages, they must be bent to a definite form for each particular row. The failure of an inner tube in boilers of this class may require the removal of several good ones.

This form of boiler has been used quite extensively and has

given satisfaction. The type shown is known as the **Speedy** type of Thornycroft boiler, another type (Daring type) being constructed with four drums; a large central upper and lower drum with a smaller lower drum on each side forming two fireboxes.

CHAPTER II.

BOILER DETAILS AND ACCESSORIES.

Tools.—The tools used by a fireman are a shovel, a slice-bar, a hoe, a rake or devil's claw, and a lazy-bar. The **slice-bar** is a heavy bar of iron with one end pointed, the other being bent into a handle. This bar is used to break clinker. The **hoe** and **rake** are of heavy construction, and are used in drawing clinker from the fire, for leveling the coal, or hauling the fire. The **lazy-bar** is a piece of iron bar bent to hook over the hinge of the door on one side, and to rest on the catch of the door on the other, the part extending over the doorway being horizontal. It is used as a support for the handle of the hoe or rake in cleaning or banking fires.

Grates.—The grates for boilers are of many forms and are designed for burning various fuels. For the hand-firing of anthracite coal the plain grate and the herring-bone, or Tupper grate, are those most often used.

Plain Grate.—This grate, Fig. 19, is formed of **bars** of

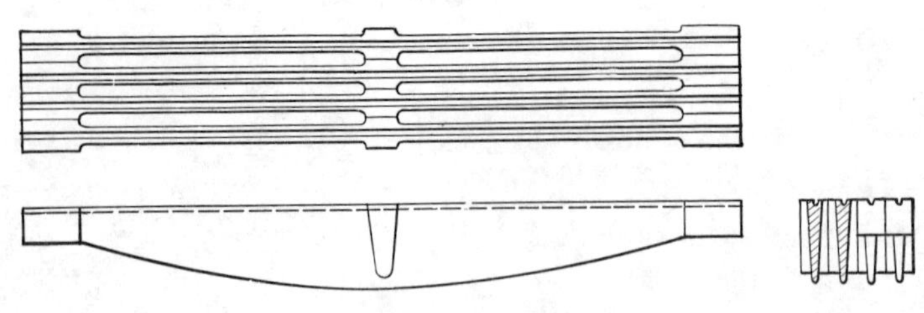

FIG. 19.—Plain Grate.

cast-iron, with **lugs** or projections at each end and at the center. They are made narrow and deep, so that with a comparatively large opening between the bars the grate is strong enough to stand the heat of the fire and the rough **handling it** receives. The projections are used to prevent the bars from warping. These bars are usually cast in pairs, and the grooves on the upper surface serve to guide the slice-bar as it is used to break the clinker. The grate shown in Fig. 20 is cast in sec-

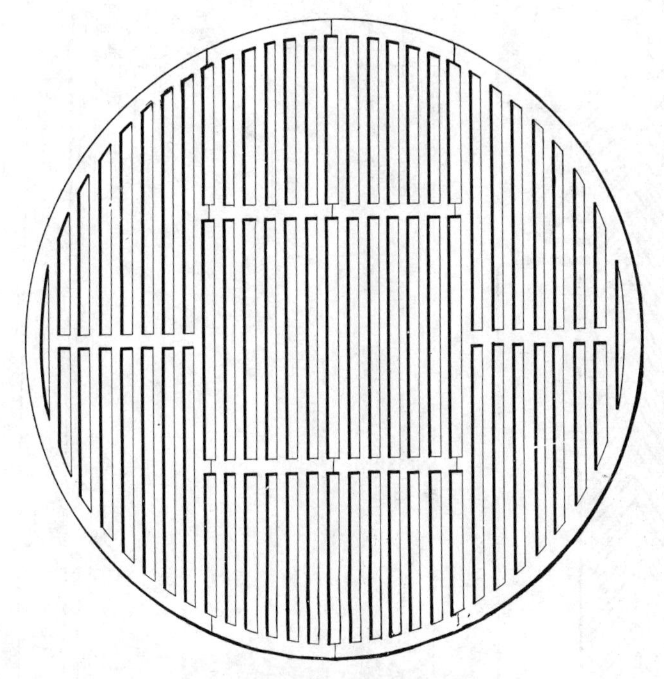

FIG. 20.—Circular Grate.

tions and conforms to the outline of the fire-box, giving a stiffer grate, but one which requires a new section when a single bar burns out.

The **herring-bone,** or **Tupper grate,** is composed of two bars united along their length by a series of narrow V-shaped bars, and at their ends by wider plates. These crosspieces, made as shown in Fig. 21, allow for expansion and contraction.

The opening between the bars, called the **air-space,** is governed by the kind of fuel, being about three-eighths of an inch for small coals such as pea or buckwheat. To secure a

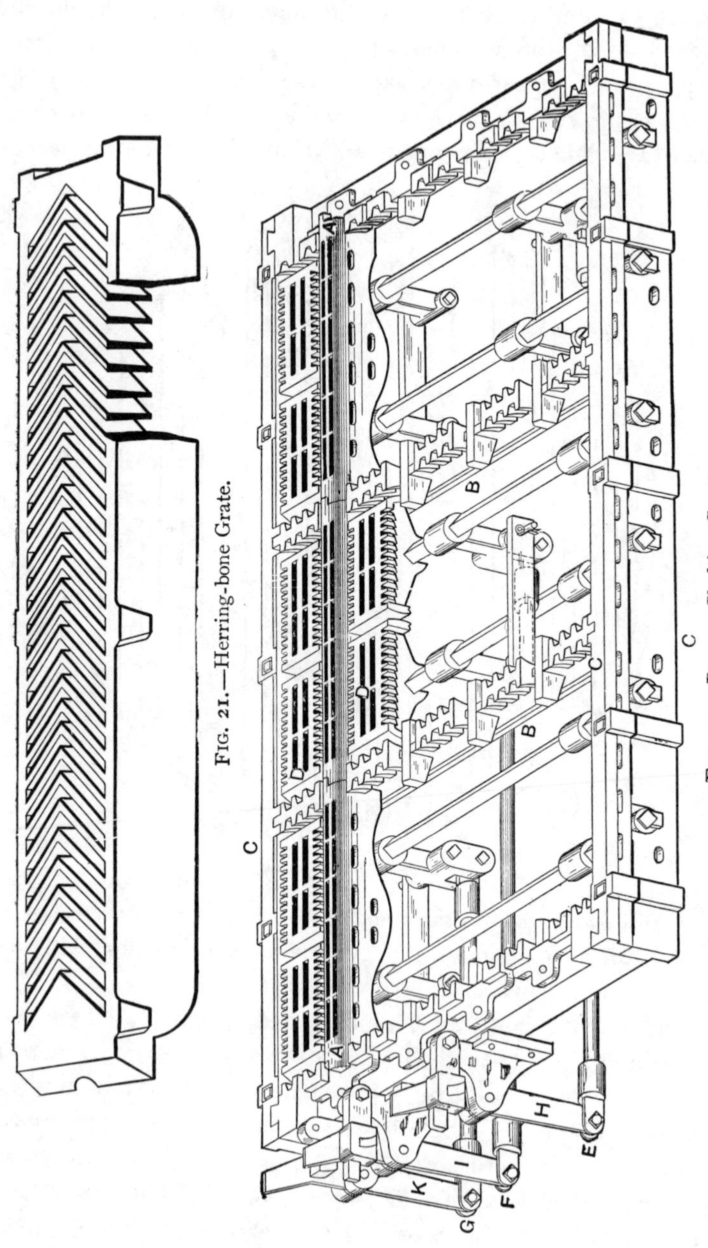

Fig. 21.—Herring-bone Grate.

Fig. 22.—Reagan Shaking Grate.

satisfactory supply of air, the total air-space should be at least fifty per cent of the grate area. The air for combustion cools the bars and prevents burning out.

Shaking-grates.—To burn soft coal rapidly, a large supply of air should be provided and the fuel should be kept well broken up on the grates.

The **Reagan grate** has been designed to meet these requirements of soft coal. The type of this grate, as adapted to the Scotch marine boiler, is shown in Fig. 22. It consists of a series of stationary bars A, called **dead-bars,** which are supported by bars B carried on the side-pieces C. Between these dead-bars are the **choppers** D, rocking on shafts supported on the side-pieces C. These choppers are moved by means of the rods E, F, and G, and the **bell-crank levers** H, I, and K, the choppers of the front section being so connected that they are all moved by K, while H and I control the other sections of the grate.

The choppers serve to break up any clinker and allow it to fall to the ash-pit, and, at the same time, to break up the fire above. The dead-bars being above the level of the choppers tend to support the coal, so that the air has free access between them. The advantages of this grate are the high rate of combustion with the complete burning of the coal, the fire being in a very porous condition, and the fact that it is very seldom necessary to clean fires in the ordinary manner. Tests for seventy-two hours have been run without cleaning the fire.

Mechanical Stokers are extensively used because they permit the use of poor fuel and decrease smoke production. There is a gradual but continuous supply of fuel, which is ignited readily, and the small but constant amount of volatile matter is ignited by the incandescent fuel below. The amount of labor is reduced, as fires on such grates are self-cleaning, and with overhead coal-bins it is an easy matter to fill the fuel-hoppers. Fuel economy is also claimed for these stokers, as there is a regular supply of coal and the opening of doors is unnecessary. The **Wilkinson stoker,** Fig. 23, is formed of a series of box-castings or grate-bars A, containing steps, and so supported at each end on a planed surface that they may be slid back and forth by the rods B. The steps are inclined at about twenty-five degrees. In the riser, or vertical face of each step, are holes through which the air for combustion enters. Coal is fed to the hopper C, and falling on the bars is pushed inward by the pusher D, which is attached to

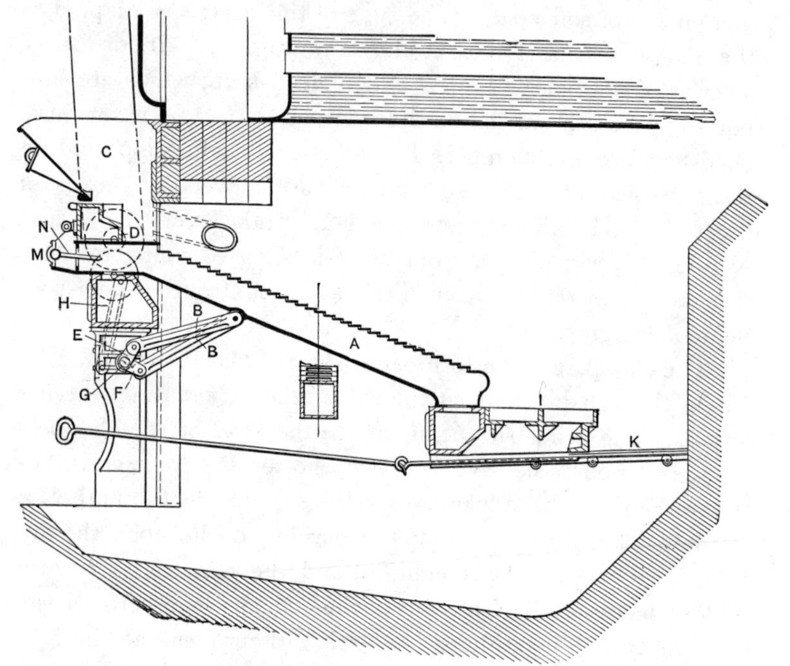

FIG. 23.—Wilkinson Stoker.

alternate bars. While one bar is moving inward the adjacent ones move outward. This is accomplished by the rods B and the arms E and F, the latter being fastened to a shaft G, oscillated by the rod H, which is driven by a system of gear-wheels. In this way the coal enters the stoker and gradually pushes the burning fuel before it, the ash finally being discharged on the grate I and slide K at the bottom. The slide K prevents the air from entering at this point and compels it to pass through the coal. The motion of the bars pushes the ashes from the grate I, as they collect. To keep the grate-bars from becoming clogged with clinker, and so being burned out, and also to aid combustion, a steam-jet is introduced in the hollow grate-bar by the steam-pipe M and nozzle N. This consumes considerable steam and is an objectionable feature for that reason.

Furnace Doors.—The fire-doors for boilers are made of cast iron or stamped metal. They must be durable, as they are subject to rough usage, and should fit so snugly when closed that no air can be drawn in. To admit air above the fire when desired, a **grid** or **register,** A, Fig. 24, is fixed **to the door.**

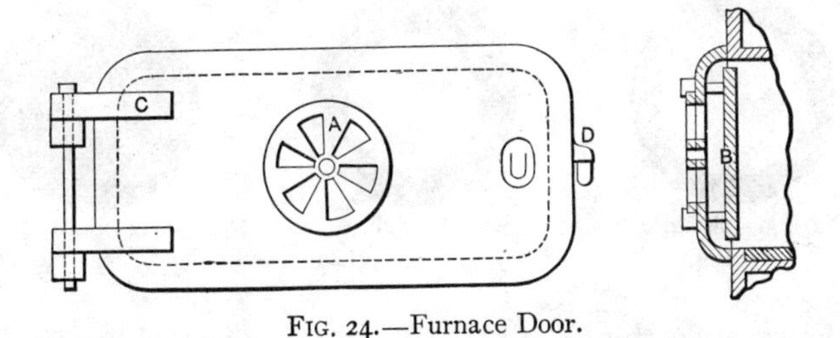

FIG. 24.—Furnace Door.

This may be closed or opened at will; but as it may be broken it is sometimes omitted from doors, and, in place of it, a number of holes are made in the casting. To protect the front part of the door from the direct action of the heat, and to better distribute the air over the fire, a baffle-plate B is attached to the casting. This plate contains small holes not in line with those in the front plate. The door is mounted on stout hinges C, and is held closed by the lug D. This lug is sometimes replaced by a latch. For marine boilers the latch, Fig. 17, is so formed that it prevents the door from moving when opened wide.

Dampers.—The dampers shown in Figs. 1, 2, 6, and 8, of Chapter I, are constructed of sheet or cast iron mounted on **trunnions** or shafts, and having attached, outside of the flue, a wheel or arm by which they can be moved. By a **counter-**

weight, or by the position of the trunions relative to the center of gravity of the plate, the damper tends to close. It is kept open by a chain or cord, which is fastened in the proper position by the fireman, or moved by an automatic damper regulator, which is controlled by the steam pressure.

Pipe.—The pipes used in connecting boilers are usually made of wrought iron or steel, although cast iron is sometimes used for water-pipe. The wrought-iron pipes are made by rolling and welding iron plates called **skelp,** and sometimes they are made from a solid piece by special processes. Pipes are made of varying thicknesses for use with different pressures. The section of a **standard** ½″ pipe is shown in Fig. 25. Fig.

FIG. 25.—Standard ½″ Pipe.

FIG. 26.—Extra Heavy ½″ Pipe.

FIG. 27.—Double Extra Heavy ½″ Pipe.

26 shows a section of **extra heavy** ½″ pipe, while Fig. 27 is a section of **double extra heavy** pipe of the same size. These last two forms are used when the pressure is exceedingly high, as in hydraulic machinery. The standard pipes are usually furnished in lengths of about 18 feet, with the ends threaded, although very large pipes are supplied without threading, and in shorter lengths. Pipes are named usually by a nominal inside diameter, which differs slightly from the actual inside diameter. The sizes vary from ⅛ inch (actual diameter, 0.27 inch) to 30 inches.

Another system of piping which is often employed for the larger size of pipes makes the nominal diameter the same as the outside diameter, and hence this is spoken of as **Outside Diameter,** or **O. D.** pipe. The standard sizes of pipe usually used in steam-engine work are ¼, ⅜, ½, by quarter inches to 1½″, by half inches to 5″, 6″, 8″, 10″, 12″.

Cast-iron Pipe is cast in lengths of 12 feet and of various thicknesses and diameters. It is seldom used in boiler plants, its principal use being for water or gas mains.

Pipe Unions.—The method of uniting wrought-iron pipe is shown in Figs. 28 to 30. Fig. 28 shows the use of a **socket**

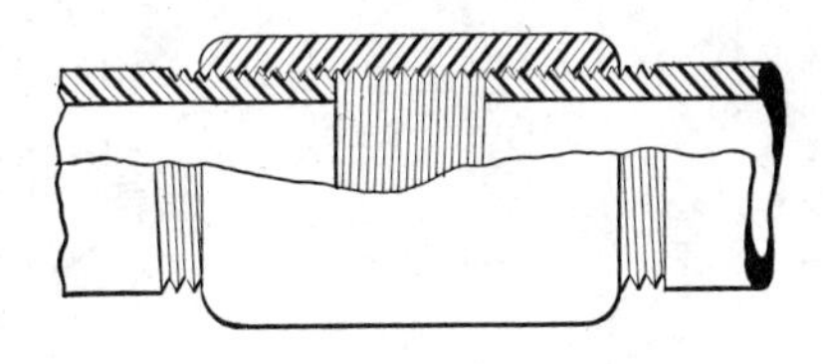

FIG. 28.—Socket or Coupling.

or **coupling.** It is a short pipe of wrought or malleable iron, threaded from each end with taper threads and into each end of which may be screwed the end of a length of pipe. When this is screwed over one pipe the next pipe must be screwed into the socket, so that in connecting up two pipes, neither of which can be turned, it is impossible to use such a coupling. In such a case it is necessary to use a coupling, one end of which is threaded right hand and the other left hand, such a coupling being called **a right and left coupling,** which can usually be distinguished from a right-hand coupling by longitudinal ribs on the outside. The turning of this coupling draws both pipes into the coupling and makes a tight joint. It is to be noted that the threads on steam-pipes and fittings are turned to a taper of $\frac{3}{4}$ inch to the foot, so that as they are screwed together the threads become tight. In addition, before screwing the pipe home the threads are painted with red lead and oil, or graphite and oil, or some patent composition, to make them steam-tight, or to permit of their easy separation.

The use of a right and left coupling requires the spreading of the pipe when making the joint; when this is impossible the **union,** Fig. 29, or **flange union,** Fig. 30, must be employed.

The union, Fig. 29, consists of two ends, A and B, and a nut C. The nut and the end A form what is known as the **female end,** and the end B is known as the **male end.** The ends of pipes are screwed into these two ends, which, after placing a

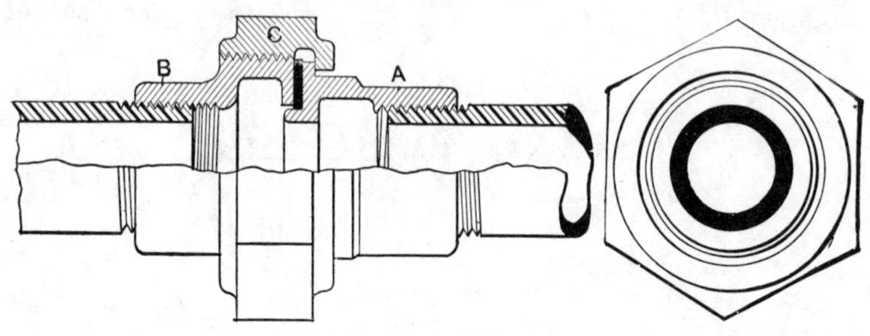

FIG. 29.—Union.

piece of canvas indurated with rubber, or some other form of gasket, between, are drawn together by the nut C. Forcing these two ends together on the gasket makes a steam-tight joint.

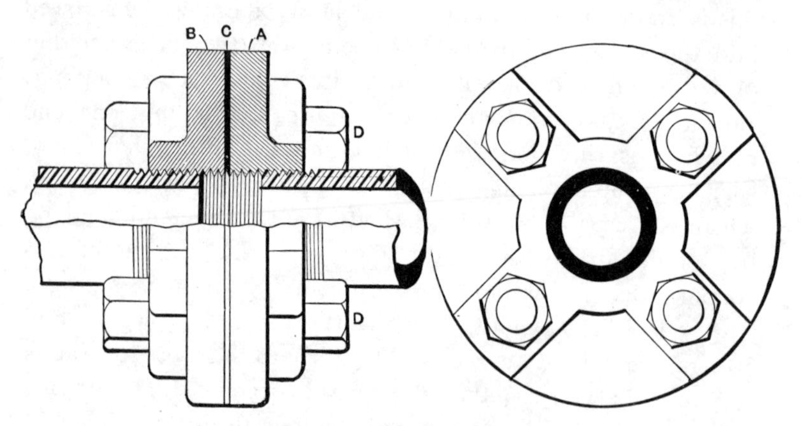

FIG. 30.—Flange Union.

The flange union, Fig. 30, is formed by two flanges, A and B, which are drawn against the gasket C by the bolts D, after being screwed to the pipe ends.

Cast-iron pipes are united by flange joints, the flanges being cast with the pipes, or by the **bell and spigot** joint shown in Fig. 31. The **bell** or **socket** A is cast on one end of the pipe, and a projection B is cast on the other, which is called the spigot end. In making the joint some packing material, such as oakum or hemp, is forced in the bottom of the space between

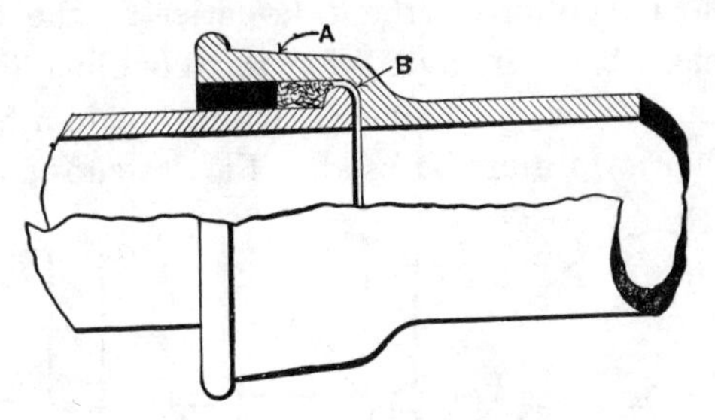

FIG. 31.—Bell and Spigot Joint.

the bell and the socket, and then hot lead is poured in, filling up the solid black space in the figure. After this lead solidifies it is driven in with a calking tool, making a tight joint.

Pipe Fittings.—To connect pipes which are at right angles, a cast-iron or malleable-iron fitting, called an **elbow or ell,** is used, Fig. 32. The pipes are screwed into each

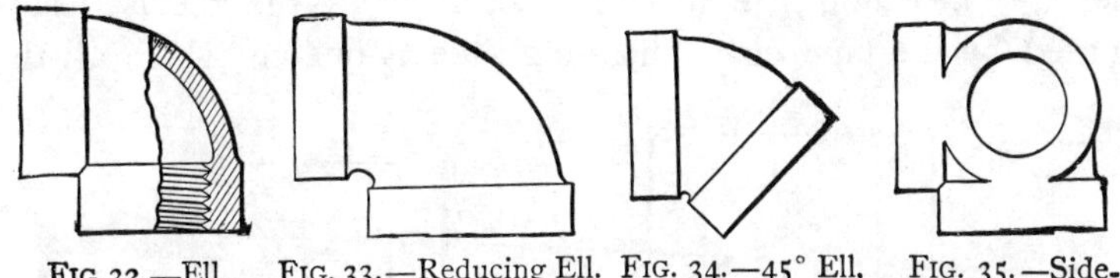

FIG. 32.—Ell. FIG. 33.—Reducing Ell. FIG. 34.—45° Ell. FIG. 35.—Side-outlet Ell.

end of this, which is threaded as shown in the figure. A **reducing ell**, Fig. 33, is used when the two pipes are of different diameters. The size of the elbow is designated by the size of the pipe, a **two-inch ell** being one tapped for two two-inch pipes. Both sizes are given for the reducing-elbow; thus a **2″ × 3″ ell** is one intended for a 2-inch and a 3-inch pipe. Fig. 34 shows a **45° elbow,** spoken of as a "45." It is used in making offsets in a pipe line, or to prevent sudden change in direction. When a branch is to be taken off at a right-angle bend, perpendicular to the two pipes, a **side-outlet elbow,** Fig. 35, is used, and when a pipe is to be returned parallel to its original direction a **return bend,** Fig. 36, is employed.

With the exception of the side-outlet ell the fittings just described have been intended for the continuation of a pipe line. When, however, it is necessary to obtain a branch from a line, another form must be used. Fig. 37 shows a **tee** which

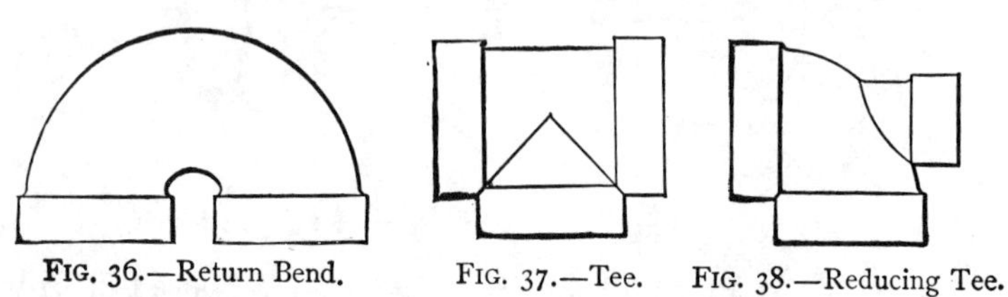

FIG. 36.—Return Bend. FIG. 37.—Tee. FIG. 38.—Reducing Tee.

is used for making a branch in a pipe line at right angles to the line. Fig. 38 is a **reducing-tee,** while Fig. 39 is a **Y** used when it is necessary to run the branch at 45° to the main line. The projecting portion of a tee is spoken of as an outlet, the remaining portion being the run. When the run and the outlet are the same size, as shown in Fig. 37, the tee is designated by the pipe size; thus, a 2″ tee is one in which all the

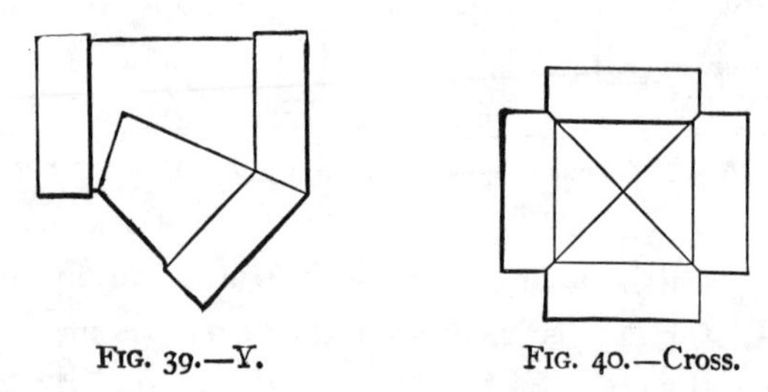

FIG. 39.—Y. FIG. 40.—Cross.

openings are for 2-inch pipes. In the case of a reducing-tee the dimensions of the run are given first and then the outlet; thus, Fig. 38 shows a $2'' \times \frac{3}{4}'' \times 2''$ tee, and by this it is known that the run reduces from 2 inches to $\frac{3}{4}$ inch and the outlet is 2 inches.

When more than one branch is desired a **cross,** Fig. 40,

or a **manifold**, Fig. 41, or a **side-outlet tee** is used, according to the requirements.

To reduce the size of a pipe line at any point two methods

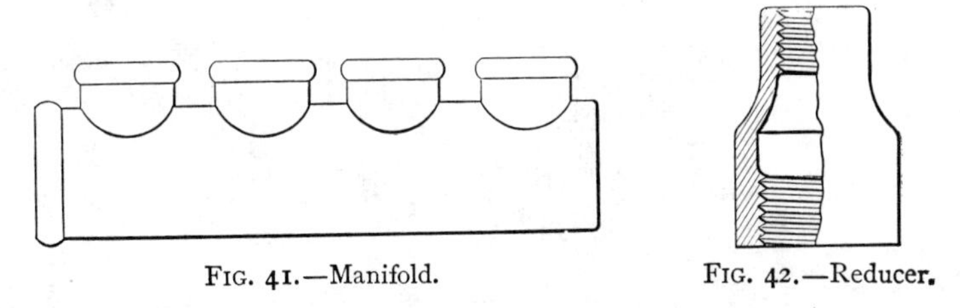

FIG. 41.—Manifold. FIG. 42.—Reducer.

are employed; either a **reducer** or **reducing-coupling,** Fig. 42, is used on the end of a length of pipe, or a **bushing,** Fig. 43,

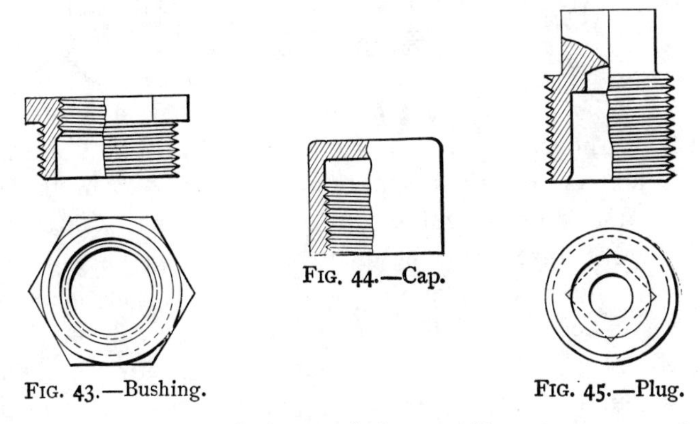

FIG. 43.—Bushing. FIG. 44.—Cap. FIG. 45.—Plug.

is screwed into some fitting and the smaller pipe screwed into the bushing or reducer. The **cap,** Fig. 44, is screwed on the end of a pipe line to close it, while the **plug,** Fig. 45, is screwed into a fitting for the same purpose.

Fittings with bends of larger radii than those shown are termed long-sweep fittings. In place of screwed fittings, as shown in Figs. 32 to 45, **flange-fittings** are often employed. These are made with faced flanges in place of the screwed ends, and to these are fastened the flanges on the pipes in a manner similar to that shown in Fig. 30. When such a pipe line is to be closed a plate called a **blank flange** is bolted to the last flange on the line and the pipe is said to be **blanked off.**

Short pieces of pipe are called nipples. These are divided into the **close nipples**, Fig. 46, the **short** or **shoulder nipple**, Fig. 47, and the **long nipple**, Fig. 48, depending on the length.

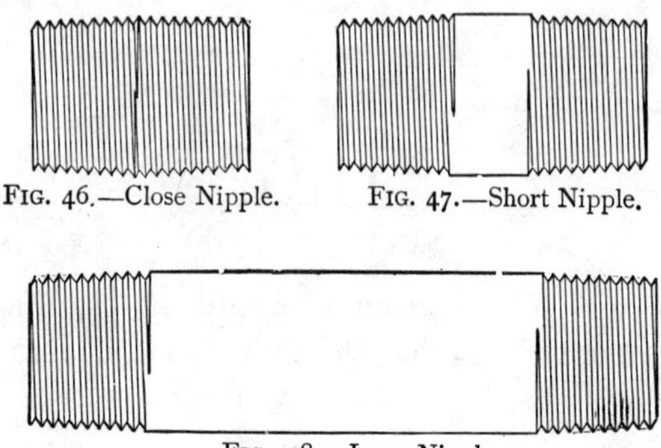

FIG. 46.—Close Nipple. FIG. 47.—Short Nipple.

FIG. 48.—Long Nipple.

Feed-line and ***Check valves***.—Where a pump is used in a boiler-house there is usually a large feed-line from which all the boilers take their supply of feed-water. The branches each contain a **check-valve** to prevent the return of water and a stop-valve to regulate the quantity. A valve should always be placed next to the boiler, so that, should the check-valve become fouled, or fail to work for any cause, it may be examined by closing the valve and shutting down the pump. The regulating-valve should be placed at a convenient point for the control of the feed. The construction of the check is shown in Fig. 49. The valve is swung from the **pin** A and rests on the **seat** B. It contains a composition disc which can be easily replaced when necessary. The arm C should be so made that the valve is free to seat itself if it should wear unevenly. To repair the valve the cap D is taken off and the **disc** F renewed. The action of this valve is clearly seen from the figure, water, entering as shown, lifts the valve readily, but when this current ceases the valve closes, preventing the return.

Stop-valves.—There are two classes of stop-valves in common use for water and steam—the globe valve and the gate-valve.

Globe Valves.—These valves, shown in Fig. 50, are made of brass throughout, or of cast iron with brass parts where corrosion is apt to occur. The spindle A passes through the removable cap B and forces the valve-disc C to its seat D. The disc is made of a rubber composition and fits the cap E, from which it can be removed when necessary. A valve can in this

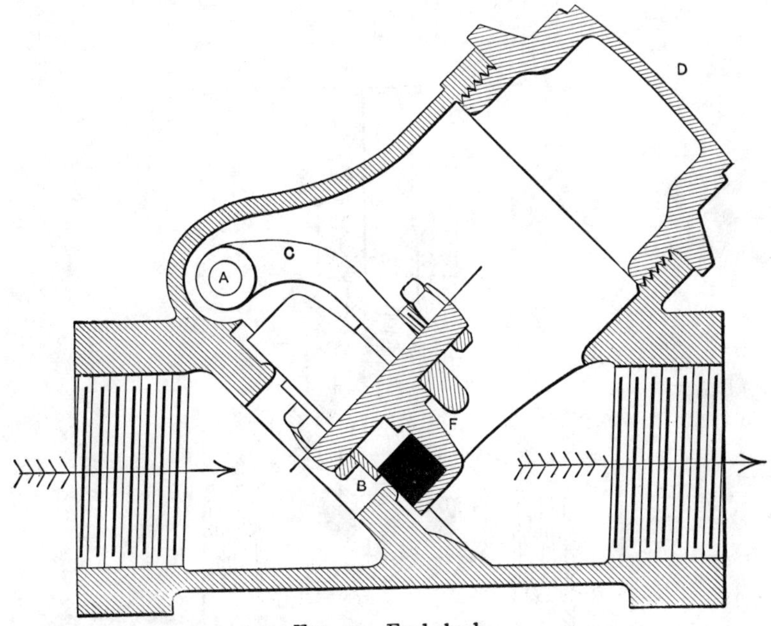

Fig. 49.—Feed-check.

way be repaired quickly and cheaply. Water or steam is prevented from leaking around the spindle by wrapping it with cotton wicking, or other packing, under the **stuffing box** F, and, on screwing this down, the packing closes in against the spindle. The valve shown in the figure is known as a **screw-ended valve,** because the pipes to which it is attached are joined by screwing them into the threads on each end. A flange-valve has the junction made by flanges, as is shown in Figs. 52 and 55.

Gate valves.—The construction of a gate-valve is shown

in Fig. 51. The body is of cast iron and the **mountings** of bronze. These include all the wearing surfaces which are subject to corrosion. The spindle A is prevented from moving vertically by the collar B, and on turning A the top wedge and nut C are raised and the pressure between the gates DD is relieved. Finally, as C is still further raised, it lifts the

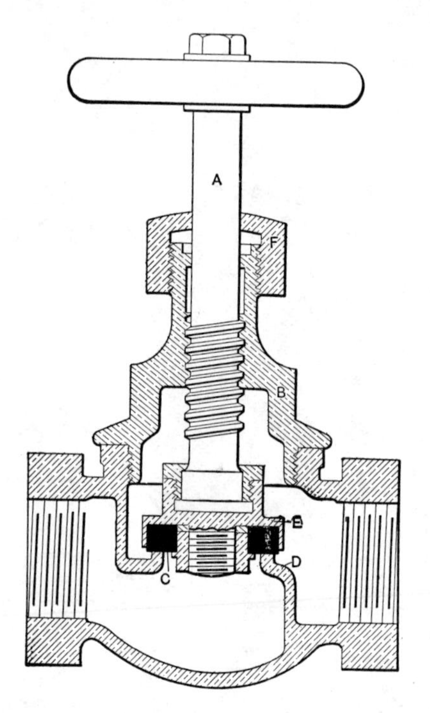

FIG. 50.—Globe Valve.

gates by projections on their backs which are not shown in the figure. Other projections on the lower wedge E come in contact at the same time with a ledge on one of the gates, and in this manner E also is raised. These parts enter the cap or **bonnet** F, leaving a straight path for the fluid On turning A in the opposite direction the gates are forced into their lower position, and on further turning, the upper wedge is driven down on the lower wedge, tightening the gates DD against

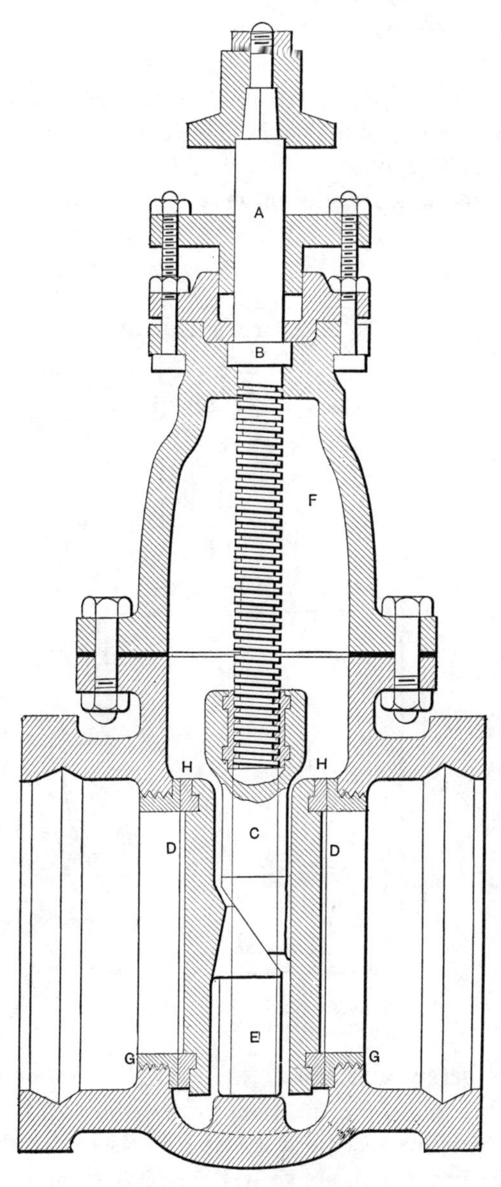

FIG. 51.—Ludlow Gate Valve.

the bronze seats GG. The gates are each faced with a removable ring H. The valve shown has bell-ends, which are used where **calked joints** are made with lead, as is the practice on water and gas mains. This valve is also spoken of as having an **inside screw** on the spindle.

Fig. 52 shows an **angle-valve** with flange connections and

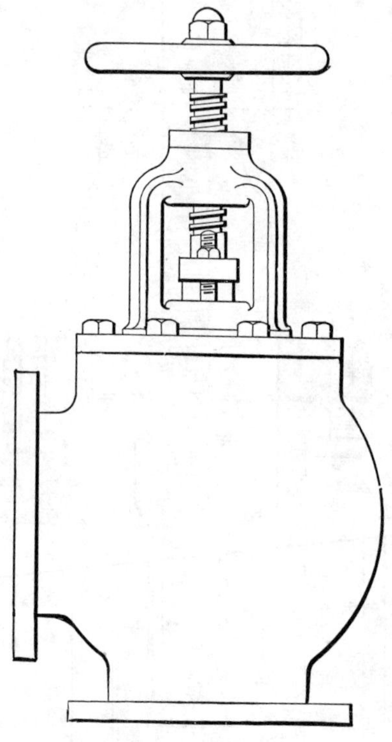

FIG. 52.—Angle Valve.

an **outside screw and yoke.** The angle form is employed where there is not sufficient room for an elbow and an ordinary valve. The valve disc in this instance rises with the spindle, while in the gate-valve above, the disc moves on the spindle.

Blow-off Valves and Cocks.—The **blow-off valves** are so designed that no pockets or obstructions are formed which would check the flow of thick fluids or muddy water, and that there may be little danger of material lodging under the seat. Their construction is shown in Fig. 53.

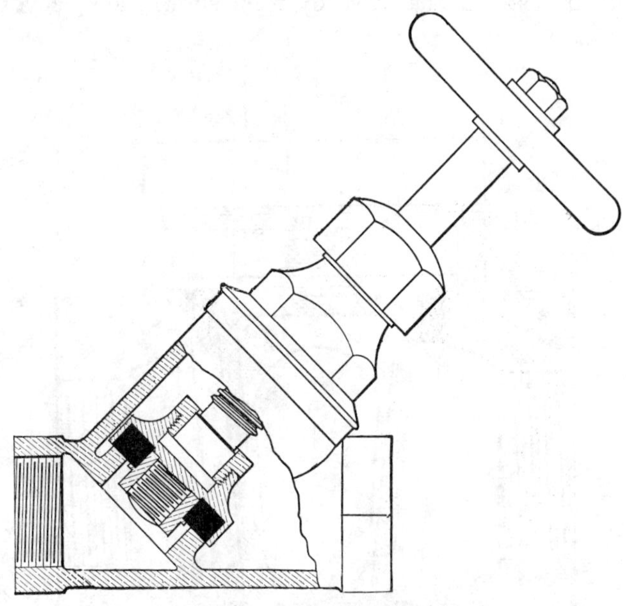

FIG. 53.—Blow-off Valve.

A blow-off cock is constructed as shown in Fig. 54. It consists of a **conical plug** E fitting in the cavity of a casing A. The plug contains an opening extending through it on a diametral line. This, in one position, corresponds to the passage in the casing and on turning the plug through a right angle into the position shown in Fig. 54, the opening comes opposite the solid wall of A and the body of the plug closes the passage. The casing is sometimes made with a removable **asbestos seat,** or bushing B, in the form of a hollow frustrum of a cone and pierced by openings CC in the direction of the casing cavity. These asbestos parts can be replaced in case of wear.

The asbestos washer D holds the plug in place and

prevents leakage around the stem, and is secured by the gland F.

Steam Outlets.—Steam may be collected through a perforated dry-pipe, as has already been shown, and leaves the

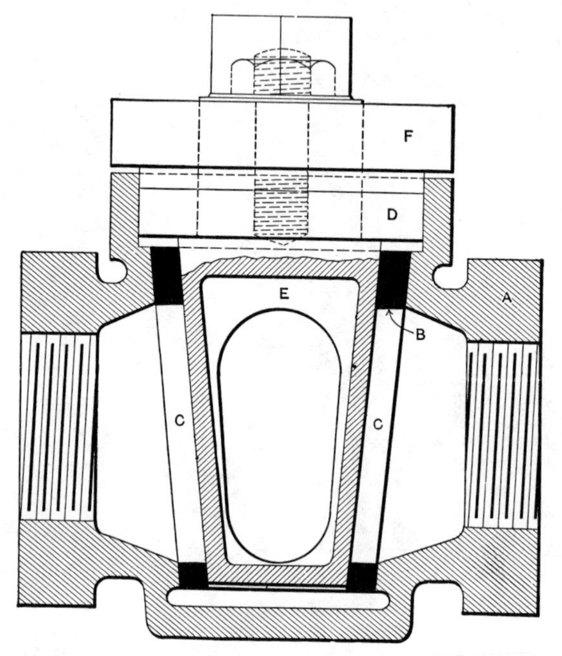

FIG. 54.—Blow-off Cock.

boiler through the steam-nozzle A and stop-valve B, Fig. 55. This valve should be placed at a high point, so that when closed no water can collect on either side. The valve B may be similar in form to any of the stop-valves previously shown, or a combination stop- and check-valve may be used. This valve, Fig. 55, allows the boiler to send steam into the line when the pressure is that of the line, but should a tube burst or the fire be drawn, the steam from the line will be automatically shut off by the valve closing. By screwing down the spindle the disc is prevented from rising and the valve acts as a stop-valve.

Safety Devices.—Safety-valves, water-columns, gauges, and fusible plugs, are intended to give warning when the correct conditions under which the boilers should work are not maintained.

Safety-valve.—There are two forms of safety-valve in common use—one the lever-valve and the other the spring or

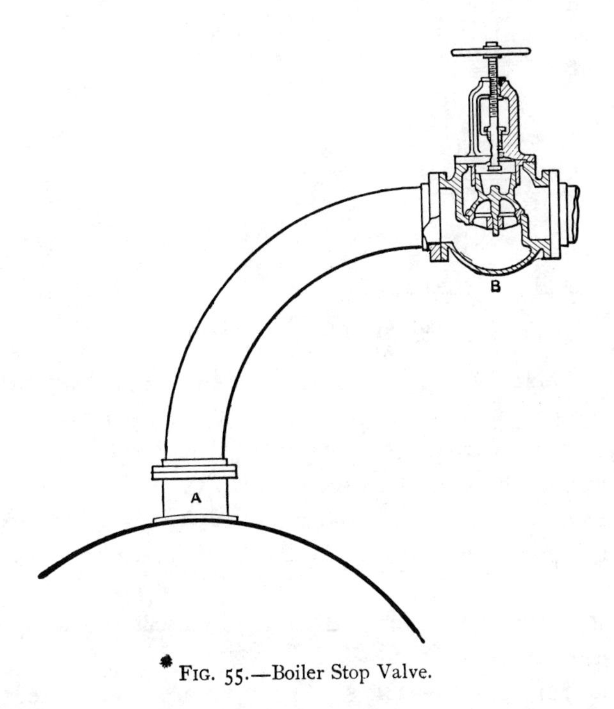

FIG. 55.—Boiler Stop Valve.

pop safety-valve. The object of each is to relieve the boiler of steam when it is not taken off by the steam-line as fast as it is generated. To accomplish this properly, the safety-valve should be of such a size that it will discharge, when necessary, the maximum quantity of steam generated by the boiler. The supply of steam being greater than the demand, the pressure rises, and, on reaching that for which the valve is set, opens it and discharges the excess steam. When the pressure has been reduced the valve closes, preventing a further loss of steam, and, should the pressure again rise, the same action would be repeated.

The **lever safety-valve** is shown in Fig. 56. The valve-

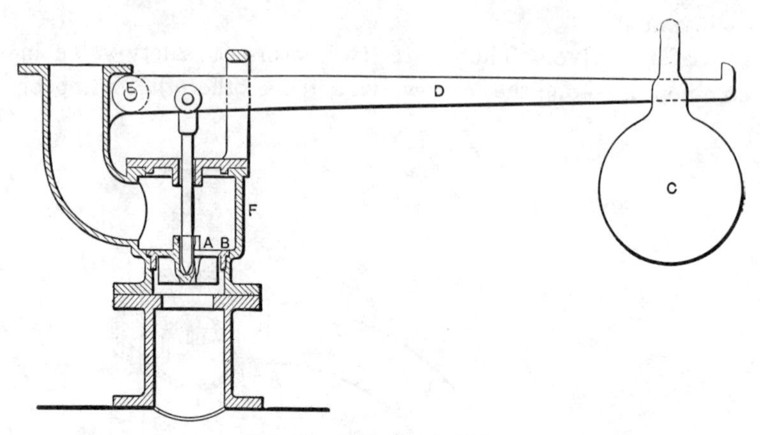

FIG. 56.—Lever Safety Valve.

disc A is held on its seat B by the weight C acting through the lever D. The seat is conical and the valve is ground to the seat to form a steam-tight joint. The lever D, which has its fulcrum at E, multiplies the effect of the weight C so that a small weight may be used. By moving the weight along the lever the pressure at which the safety-valve will "**blow off**" is changed. The body of the valve F is of cast iron, but on account of corrosion the valve-disc and its seat should be made of bronze.

The **pop safety-valves**, Fig. 57, are constructed with springs in place of levers and weights. The **spring** A is encased in a metal box B, to protect it from the discharging steam. It is forced downward by the loose collar C, which is acted on by the nut D. The spring is supported at its lower end by the collar E, on the spindle F. The compression produced by screwing down the nut D is transferred by the spindle to the valve-disc G at a point below the seat. This method of attaching the spindle to the valve causes the latter to seat fairly after blowing off. To prevent the pressure from being increased by tightening the nut, a plug H fitting a slot on the nut D is locked on the cap. The spindle may be raised by means of the lever I, thus allowing the valve to blow off at a lower pressure. This operation should be performed once each day to insure the free action of the valve. The valve-disc shown has a grooved lip projecting over the seat.

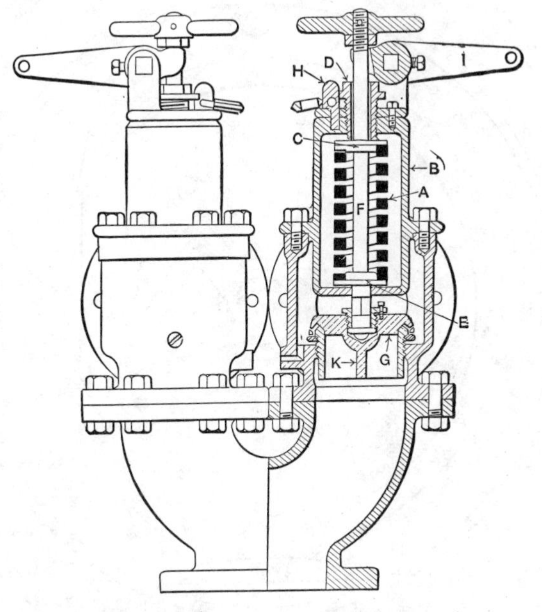

FIG. 57.—Crosby Pop Safety Valve.

This causes the steam as it leaves to be deflected downward, and thus produces an additional pressure which keeps the valve open until the boiler pressure has been reduced three or four pounds; the valve then closes instantly. The valve is guided in its movement by wing-guides K.

Gauges.—A **Bourdon pressure gauge,** shown in Fig. 58, is used to indicate the steam pressure. The tube A is formed from a circular tube which has been flattened until the cross-section is similar to that shown in Fig. 59. It is then bent into an arc of a circle and the end attachments brazed on. It the tube were circular in section the application of an internal pressure would not cause any movement of the ends. If the tube has a flattened section and pressure is applied, the ends will move outward to a larger radius.

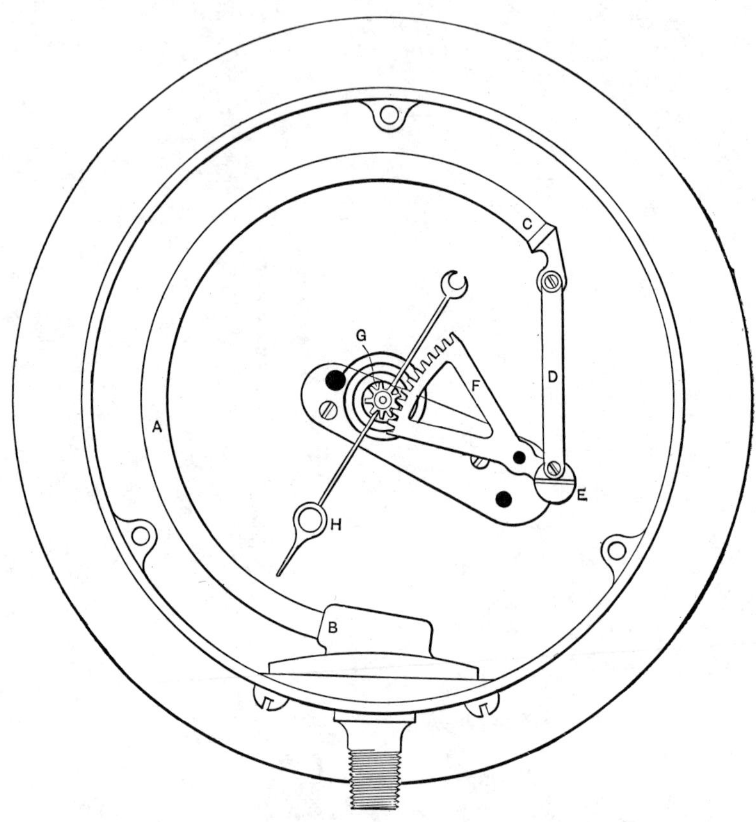

FIG. 58.—Bourdon Gauge.

The action of such a tube can be seen from a consideration of a portion of the tube **adcb,** shown in Fig. 59. The lines **ad** and **bc** represent cross-sections of the tube perpendicular to the center line. The intersection of these lines determines the center O of the circle to which the tube has been bent. If internal pressure be applied to the tube, the flattened surfaces open out, **ab** taking the position **a′b′** and **cd** the position **c′d′.**

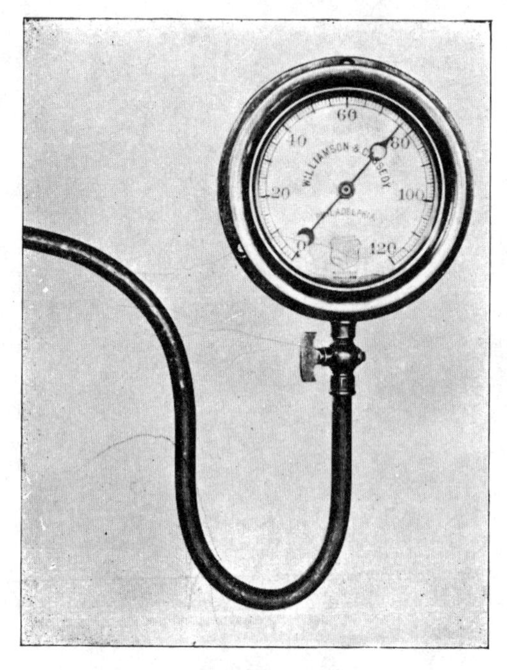

FIG. 61.—Gauge and Siphon.

(*To face page* 55.)

These lengths are not materially changed by this action, and **a′d′** and **b′c′** are the new positions of the sections considered. Continuing these lines gives the new center O′ of the arc of this portion under the new conditions. Since the section considered is any section, the action is the same for the whole tube, and as the radii are larger, the free end of the tube must have moved outward, as shown in Fig. 60.

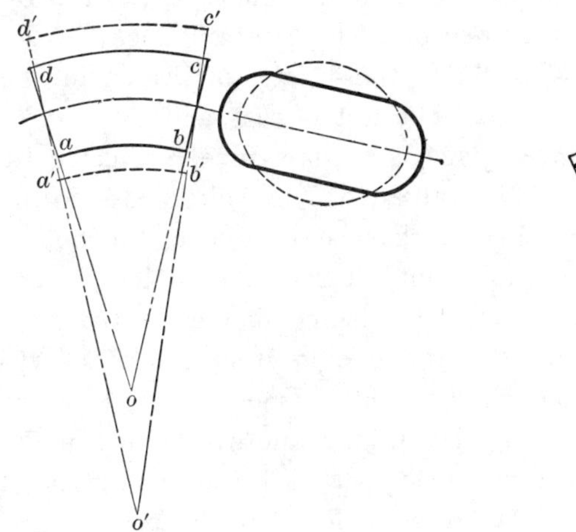

FIG. 59.—Bourdon Tube.

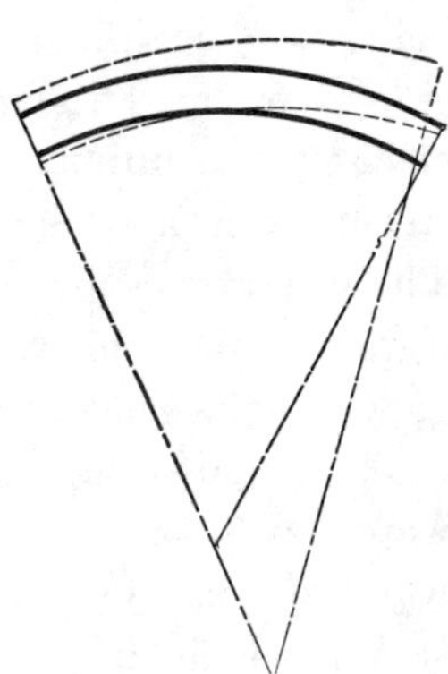

FIG. 60.—Bourdon Tube.

The end B of the bent tube, Fig. 58, is fixed, and the motion of the other end C is transmitted by link D to an arm E of a **sector** F. The teeth of the sector mesh with those of a small **pinion** G, to the shaft of which the **pointer** H is attached. A small **helical spring** is attached to this **shaft,** or **arbor,** so that the pinion is always held against the sector. The pointer H moves over a **scale** graduated to read pounds per square inch above the atmosphere, inches of mercury below the atmosphere, or any other unit of pressure which may be employed. In connecting the gauge to a steam-pipe a small cock or valve should be placed beneath the gauge, and to this a bent pipe leading to the steam line, Fig. 61. The object of the bent pipe, or **siphon**, is to prevent the steam from acting directly on the gauge, thus heating it and changing its readings. The steam which might enter the siphon is condensed and lies there as water protecting the gauge. Gauges are sometimes made with a Bourdon tube on each side, but the principle is the same as that of the single tube. **Mercury gauges** which are constructed on the principle of the hydrostatic balance, and diaphragms on which the steam produces a measured deflection, are used for steam-gauges. **Recording gauges** are made by which the pressure is recorded on a chart for reference.

Water-columns.—To show the height of water in the boiler two devices are used—the water-glass and the try, or gauge-cock. The **water-glass** is a glass tube connected by **stuffing-box unions** to two valves, one of which leads to the steam-space and the other to the water-space of the boiler. The water rises in this tube until it reaches the level of the water in the boiler. As there is some danger of the glass gauge giving incorrect readings, due to stoppage, three **try-cocks** are also supplied with boilers. One, of these is at the water-level, one above the level, and another below, in the water-space. On opening the upper cock we should obtain steam which is usually colorless close to the cock. The middle cock gives white steam, showing the presence of water, and the lower gives water, part of which becomes steam after leaving the cock. The difference in sound made by the escaping steam or water is a surer indication of the presence of water. It is now customary to place these devices on an iron casting, Fig. 62, called a water-column, and connect this by means of pipes of large size to two points of the boiler, one in the steam-space and one in the water-space.

To clean out the water-glass connections a small cock A is placed on the bottom of the lower valve casing, and on opening this, if the lower valve be closed, the upper connection of the glass is cleaned. On closing the upper valve and opening the lower, the other connection is blown free from sediment. The mud is blown from the lower end of the column through the pipe B and a controlling-valve. The gauge cocks should be tried frequently to ascertain whether or not the column is indicating properly.

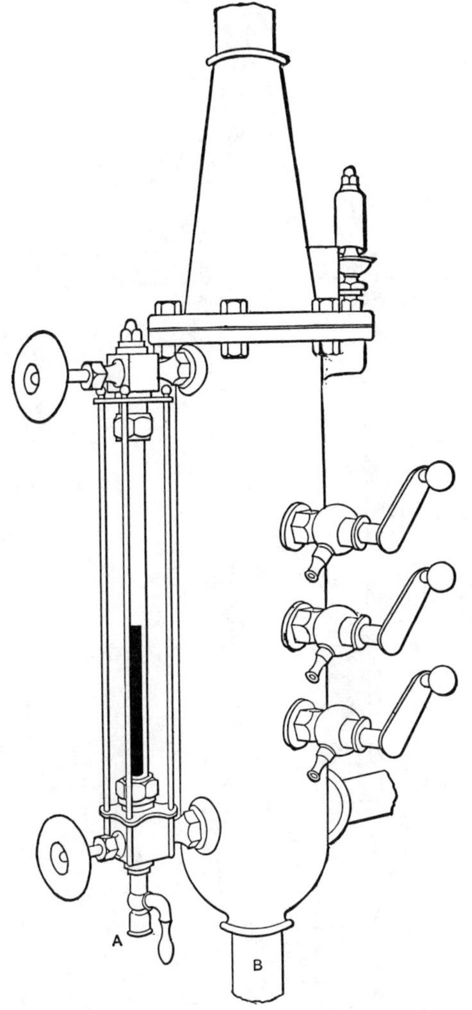

FIG. 62.—Reliance Water-column.

Water-columns are sometimes fitted with a **high and low water alarm**, one type of which is shown in Figs. 62 and 63. The column contains a copper ball or **float,** which is supported by the water. To this is joined a rod with two small adjustable collars A attached. When the water rises, the lower

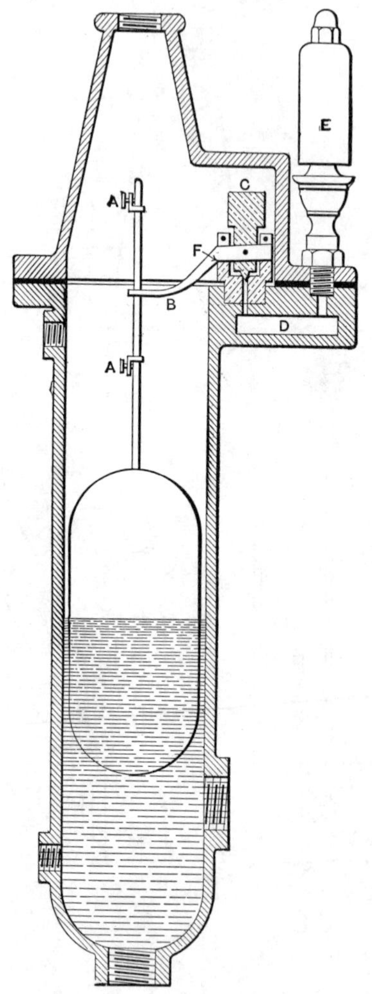

FIG. 63.—High and Low Water Alarm.

collar strikes the arm B, lifting it about its right-hand end, thus opening the valve C, admitting steam through D to whistle E. When the upper collar strikes B, as the water falls, the arm moves about F as a fulcrum, and raises the valve. The blowing of the whistle notifies the attendant of the height of the water in the boiler, and it is then his duty to regulate the feed properly. With proper management the whistle is never blown.

Fusible Plugs.—Another device used to give warning of low water is the fusible plug, Fig. 64. The main body of the

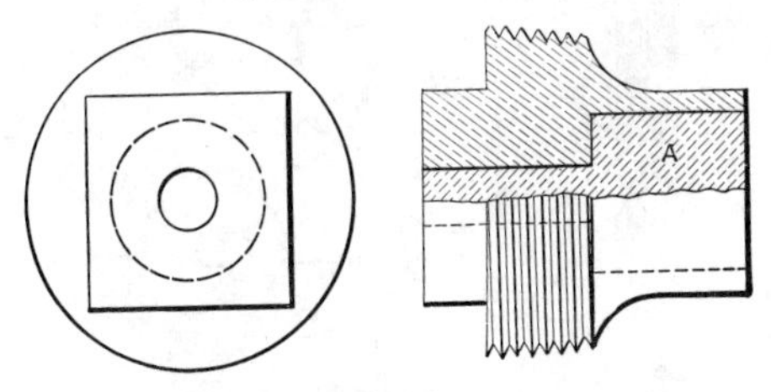

FIG. 64.—Fusible Plug.

plug is of brass with a square head and a threaded barrel. The central portion A is filled with a soft metal which melts at a point above the temperature of steam, but much below the melting-point of brass. The metal fills a hole so arranged that there is no tendency for the steam to force the soft-metal core from the body of the plug. When water is over the plug, heat is conducted away so rapidly that the temperature cannot rise, but, when not covered, the heat melts the fusible metal and the escaping steam gives warning. The plugs are placed in locations which are exposed to the heated gases, and to which little damage can be done by the heat should the water fall below their level. In this way the plug will give warning before the water sinks to a dangerous point. Care should be taken that these plugs are never covered with scale of any kind that may prevent them blowing after fusing. The locations for these plugs are in the crown-sheet of a locomotive or vertical boiler, in the rear head of a return tubular boiler, just above the tubes, and in the bottom of a steam-drum of a water-tube boiler.

Riveted Joints.—Boiler-plates are joined together by means of rivets passing through holes which have been punched or drilled in the plates. According to the arrangement of plates the joints are either **lap-joints** or **butt-joints**. The former are made by lapping one plate over the other and passing the rivets through, as shown in Fig. 65, the upper repre-

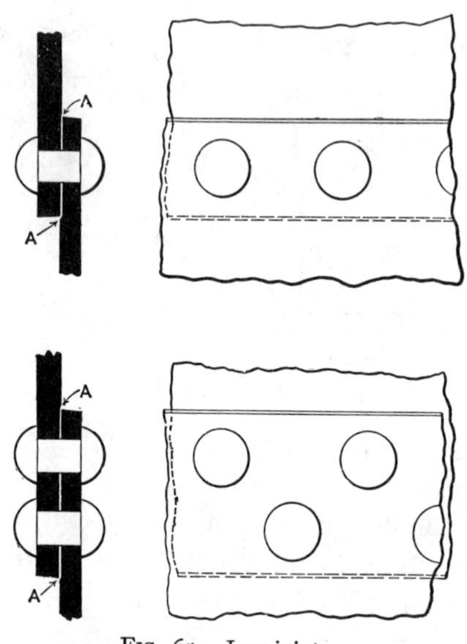

FIG. 65.—Lap-joints.

senting a single-riveted lap-joint, the lower a double-riveted one. The rivets, having a head on one end, are heated, and after being pushed through the holes in the plates have a head formed on the other end by means of hammers or **riveting machines.** The rivet on cooling pulls the plates tightly together, and, after the edge of the upper plate has been forced against the lower plate at points AA by the proper tools, this joint is steam-tight. This last operation is called **calking.** The lap-joint is usually employed to unite successive rings of a boiler shell together, forming what is known as the **ring-seam** or the **transverse** or **circumferential joint.** It is also used for uniting the fire-box and the combustion chamber-sheets of locomotive or marine boilers.

The **butt-joint**, as its name indicates, is one in which the plates butt against each other. The plates are then held together by the **butt-straps,** or **cover-plates,** AA in Fig. 66.

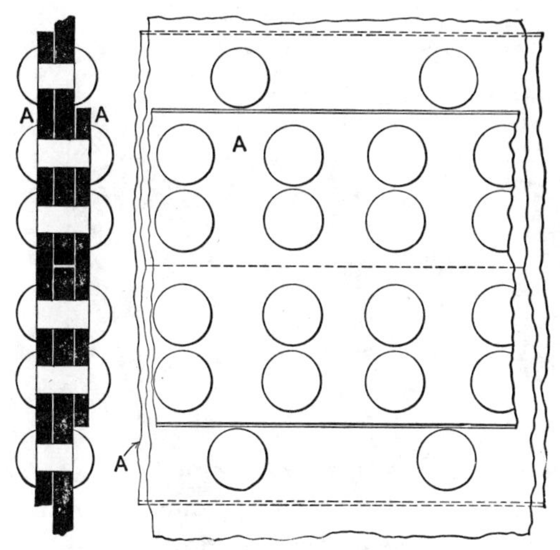

FIG. 66.—Butt-joint.

The joint is made and calked as described above. It is used in joining the two ends of a plate together in forming a ring, and this joint, because it extends along the boiler, is called the **longitudinal joint.** The butt form of joint is of great strength, and is used for the longitudinal seams because these are subject to greater stresses than the other joints of the boiler. Fig. 67 shows how these joints are arranged when they overlap. It will be noticed that the upper butt-strap on the left is flattened out, or **scarfed,** so that it extends beneath the outer plate of the lap-joint. The inside butt-straps are made wider than the outside straps, so that an extra row of rivets may be used of greater **pitch** (the distance between rivets), to make the joint stronger.

Staying for Flues.—A vessel subject to internal pressure tends to become cylindrical, and hence there is no need of stays to prevent the distortion of cylindrical shells. When a cylinder, circular in section, is subject to external hydrostatic pressure there is also no tendency to distortion; but should this be distorted a small amount, or be not truly cylindrical, the pressure tends to collapse it. Hence, when we use large furnace flues we must stay them in some manner. A very com-

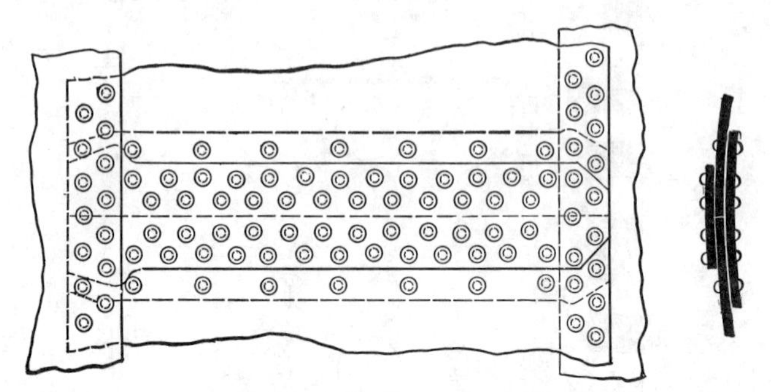

FIG. 67.—Overlapping Joints.

mon method is by corrugating the metal, as shown at A in Fig. 68. This stiffens the flue, giving a certain amount of rigidity, the corrugations acting as girders.

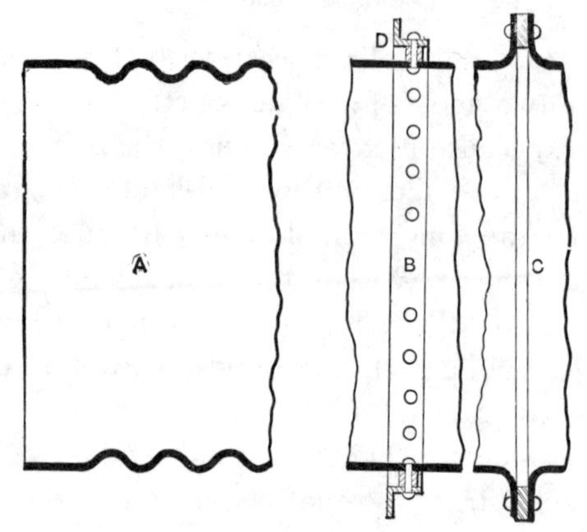

FIG. 68.—Furnace Flues.

A plain cylindrical flue is made in short sections which are flanged at the ends, and between the flanges is placed a heavier ring of metal. These, as shown at C, in Fig. 68, are then riveted together, forming what is known as an **Adamson Ring.** Instead of this, an angle-iron is sometimes riveted to the flue, as shown at B in Fig. 68. To allow water to reach the flue the angle-iron is held at a definite distance by means of **thimbles** D, of metal, through which the rivets pass.

The Galloway Tube, Fig. 5, acts as a stay for the large flue used in the Galloway boiler. The tube is made conical, so that the upper flange will cover a hole through which the lower flange can pass. The tube after being welded is flanged over and then riveted to the flue sheets.

Stays for Sheets. **Stay-bolts.**—These stays, Fig. 69,

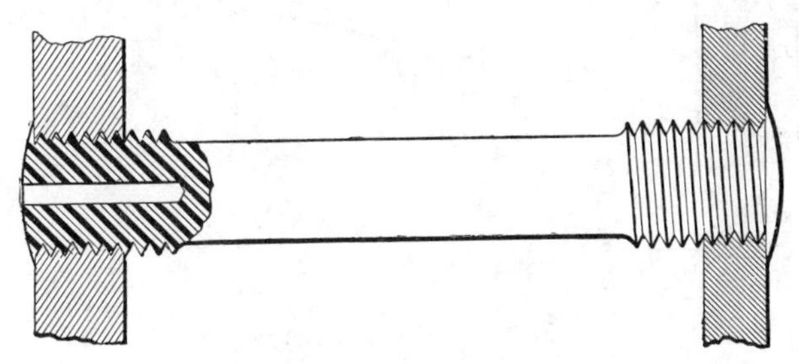

FIG. 69.—Safety Stay-bolt.

are used to stay parallel sheets such as form the sides of water-legs. They are bolts about six inches long, which, after threading from end to end, have the central threads turned off to increase their strength. These stays after being screwed in have their ends riveted over, thus making them steam-tight. They sometimes contain small axial holes extending from the outside to a point beyond the threaded portion. As stays usually break near the outside sheet, this hole allows steam or water to escape when one breaks, thus calling attention to the broken bolt. Such stays are called **safety-stays.**

Radial Stays.—The stays used above the crown-sheet of the locomotive boiler in Fig. 13 are called radial stays.

Through-stays.—When the stay extends from one end of the boiler to the other, as in Fig. 17, or from one side-sheet to the opposite one, as in Fig. 14, it is called a through-stay. Such stays are usually threaded at the ends and have nuts and washers on each side of the plate to stiffen it. The end of the stay is **upset** (made of larger diameter), so that the diameter at the **root** of the thread is at least that of the main portion of the rod.

Diagonal Stays.—When it is desired to have the central portion of the shell free from stays, the end-sheets are stayed to the shell instead of to the opposite end by means of diagonal stays. The stays are constructed in various ways, but in principle are similar. To the head, Fig. 70, is riveted a tee-

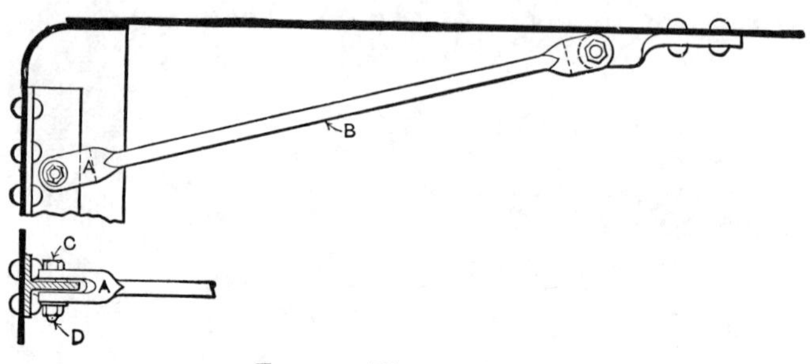

FIG. 70.—Diagonal Stay.

iron, and to it the **forked end** A of the stay B is fastened by the **stay-pin** C. At the other end the rod is flattened out and riveted to the shell, or attached to a separate piece, as shown in the figure. A flattened end is usually spoken of as a **crow-foot.** In this manner the pressure which tends to bulge the head outward is transmitted to the shell. The stay-pin is held in place by a nut, which is prevented from backing off by a **split-pin** D. Sometimes a split-pin is used in place of the nut. In some instances the end of the stay-rod is flattened and held between angle-irons. The pin form of end is sometimes used for through-stays instead of nuts and washers.

Girder Stays.—To support the top of a combustion chamber of a Scotch marine boiler, girder-stays are used. These stays, Fig. 71, are formed of two plates AA, bolted together with **distance-pieces** or thimbles B on the bolts to separate the plates a definite amount. The plates forming the girder rest only on the side-sheets of the combustion chamber, so that water can be in contact with as much of the top-sheet as possible. The loads from the stay-bolts CC are carried by the girder. These bolts are screwed into the crown-sheet and have nuts beneath, which, screwing against the sheet, stiffen the plate as well as make a tight joint. A thimble is placed between the girder and crown-sheet to prevent the sheet being distorted when the nut on the upper end is tightened, and also

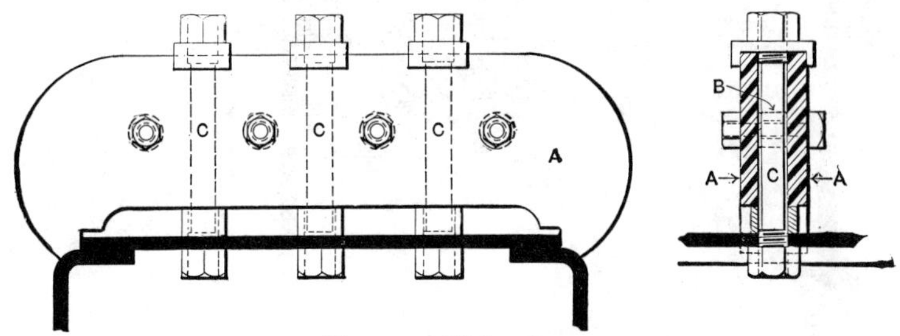

FIG. 71.—Girder Stay.

to make the girder and plate stiffer. The distance-pieces between the plates of the girder also serve the same purpose. When a girder-stay is used in locomotive practice it is so long and the load is so great that it is furnished with supports between the ends. The supports are fastened to the shell over the fire-box by sling-stays, Fig. 72. The girder-stay on a locomotive is called a crown-bar, and is formed of a special heavy "tee"-iron. To the crown-bars are attached the short-stay bolts, with thimbles to keep the crown-bar the proper distance from the sheet. The sling-stays AA are shown attached to tee-irons riveted to the shell. They are formed of two plates, one on each side of the tees, and fastened to them by the stay-pins BB.

Gusset-stay.—In the Galloway Boiler is found another method of staying the heads. This is by the use of a gusset stay, which consists of a metal plate A, Fig. 73, extending from two angle-irons on the head to two on the shell, being riveted at each end. This form of stay makes a very rigid connection, and for that reason is not used as extensively as the other types mentioned above.

Tubes.—The boiler-tubes act as stays, and where used no other stays are necessary for that portion of the tube-sheet.

These tubes are generally made seamless when of steel or brass and welded when of iron. In the Scotch marine boiler it is

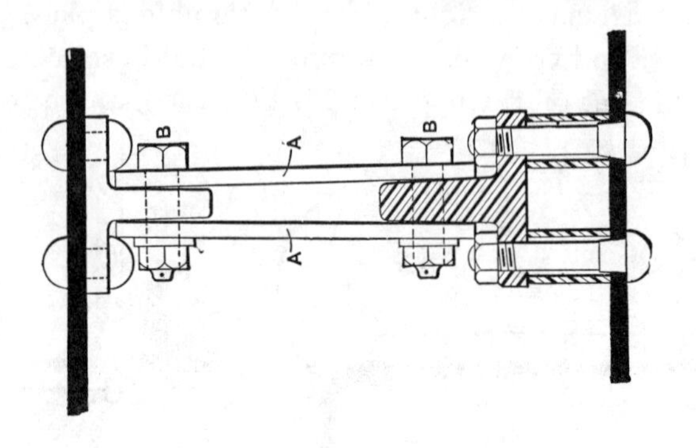

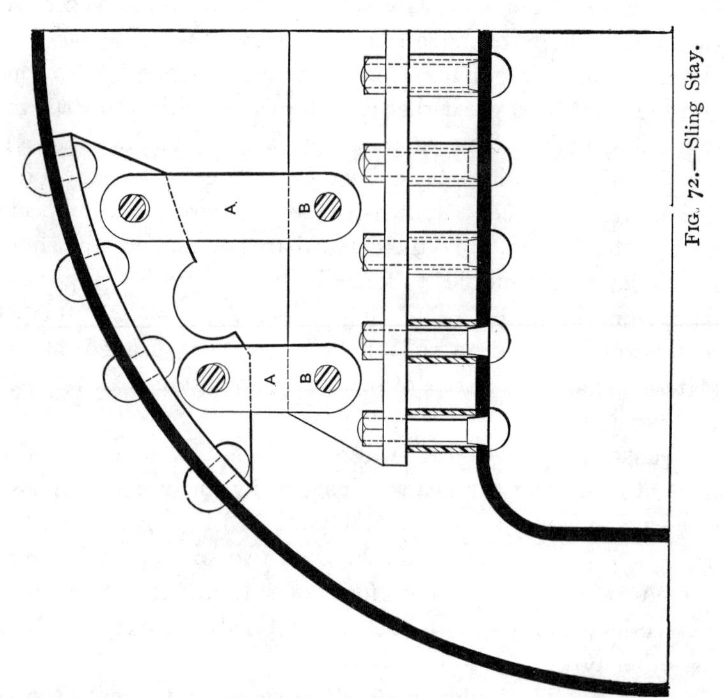

FIG. 72.—Sling Stay.

customary to make certain of the tubes, called **stay tubes,** heavier than the others, and to upset and thread their ends.

These tubes, shown at A, Fig. 74, are screwed into the plate, and sometimes nuts are used over the ends of these stay-tubes,

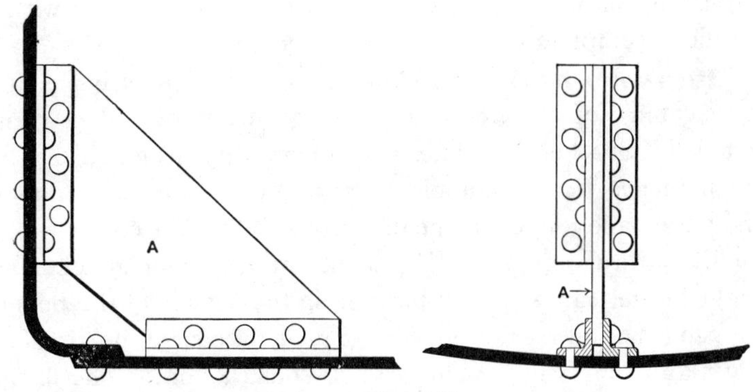

Fig. 73.—Gusset Stay.

giving the tubes a stronger hold on the plate. The ordinary tubes B of Fig. 74, are held to the tube-sheet by expanding

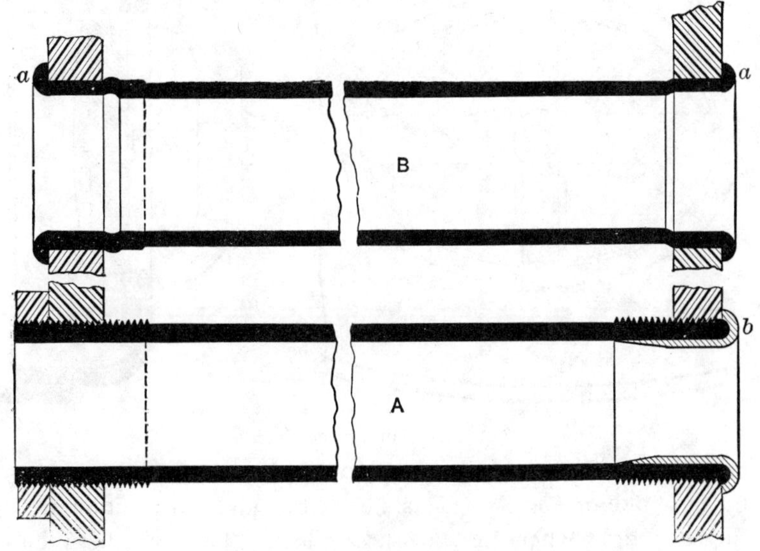

Fig. 74.—Boiler Tubes and Ferrules.

against the plate, using a special tool called an **expander.** After this operation they are **beaded** over by another tool forming the end shown at *a*, Fig. 74.

To protect the ends of tubes against the intense action of the flame, **ferrules** *b*, Fig. 74, are used. These are pieces of malleable iron with a projecting lip, which, when driven within the tubes completely cover the exposed ends.

Manholes and Handholes.—For the proper inspection and cleaning of a boiler oval openings are provided as mentioned in Chapter I. These openings vary from eleven by fifteen inches for a manhole to four by six for a handhole. They are made longer in one direction, so that the cover, which usually has a flange projecting about one inch over the opening on the inside, can be put in place from the outside of the boiler. If a cap is bolted on the outside, it may, of course, be circular. The **manhole** in Fig. 75 shows their construction. The flange

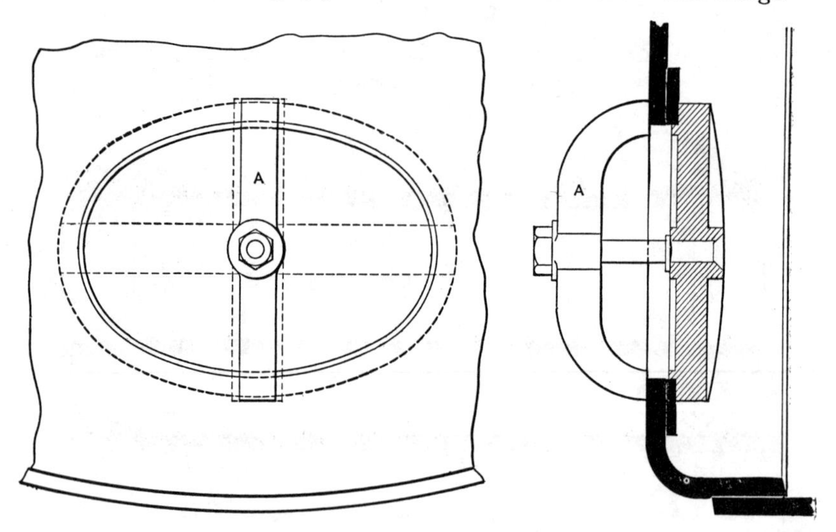

FIG. 75.—Manhole and Cover.

of the cover is held against the plate by the steam pressure and by the **yoke** or **dog** A. This keeps the plate in position and the joint tight when the pressure is low. The joint between the plate and the cover is kept tight by a piece of sheet-rubber with cloth insertion, called a gasket. Gaskets are also made of metal, of asbestos, and various compounds. Handholes are of similar construction.

Boiler-supports.—To carry a return tubular boiler, the common method is to use cast-iron brackets, Fig. 76, which,

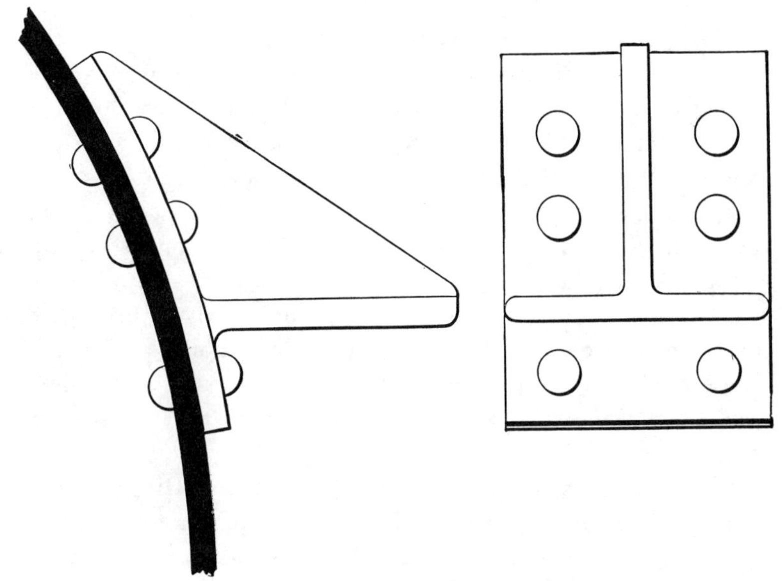

FIG. 76.—Bracket.

riveted to the shell, transmit the weight to the side-walls. This is not always advisable, as the expansion and contraction of the boiler is apt to crack the setting, unless the back brackets are carried on rollers. A better way is to support the brackets on iron columns which are properly braced, but free from the setting. The bracket has a large web to stiffen it, and should be made of tough iron.

As shown in Fig. 8, the water-tube boilers are usually carried by straps passing around the drums to a girder, which is carried on iron columns. This forms a good support and the setting is not liable to crack as the shell expands. Small boilers are carried by rings and hooks, a ring being riveted to the shell, and this fastened by a hook to an overhead beam. The methods of supporting the Cahall, the locomotive, and the marine boilers were described in Chapter I.

Although it is proper to have the brickwork free from the

boiler, it is always well to have the masonry so arranged that, should any support break, the boiler will be carried by the walls or an auxiliary pier.

Brickwork and Other Masonry.—Wherever the fire is intense fire-bricks are used. These bricks are made of a special clay which will withstand the action of gases at high temperatures. The bricks are bound together with fire-clay, and only the burned edges of the brick should be exposed to the fire. That is, wherever a brick has to be cut, its cut edge should never be exposed, as this is soft and is quickly eroded by the flame.

For the remaining part of the setting, hard red bricks are used which are bonded together with lime mortar. Some of these bricks, of a very hard texture, have been used for flue-lining and are giving fair results, even though exposed to high temperatures.

For the foundations of boilers and boiler-settings stone-masonry or **concrete** is used. Concrete is a mixture of cement, sand, and broken stone; and, when properly made, is very strong, acting as one large stone under the entire boiler.

To prevent the walls from spreading, vertical cast-iron beams, called **buckstaves,** are placed on the outside of each of the walls, and are held together at the top and bottom by stay-rods. The form of the buckstave is shown in Fig. 77. It is

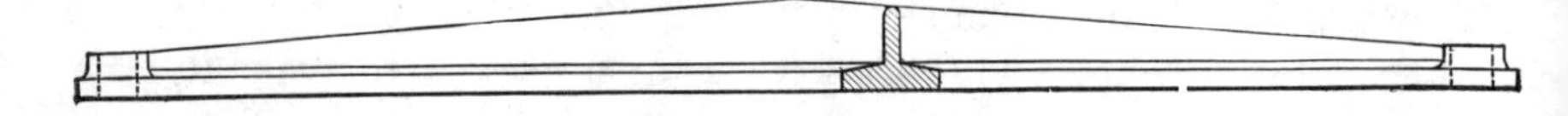

FIG. 77.—Buckstave.

a cast-iron plate with a rib on the back to stiffen it. The boilers in which columns and overhead beams are used are not built with buckstaves, as the columns answer the same purpose. The method of using buckstaves is clearly shown in Fig. 6.

Fig. 78.—Section of Single Cylinder Pump.

(*To face page* 71.)

CHAPTER III.

BOILER ROOM AUXILIARIES.

Pump.—The feed-water is forced into the boiler by a pump or by an injector. The **steam-pump,** although not efficient as a steam user if the exhaust-steam cannot be utilized, is so reliable and requires so little attention that its use is almost universal. A pump is shown in Figs. 78 and 79. A is the **steam-cylinder** and B the water-cylinder, made of cast iron. Within these cylinders are the **pistons** C and D, which fit tightly against the walls, being connected by the piston or pump-rod E. Suppose the right-hand end of A to be open to the air, and the left-hand end to the steam-pipe, the steam will push the piston to the right, driving the water from the right-hand end of B, through the valve F into the feed-line. While the piston D moves to the right a partial vacuum is formed behind the piston, and water is drawn through G from the source of supply. When the pistons reach the extreme right, or, **the end of the stroke,** this operation is reversed by the valves H and I * and the piston is driven backward to the left, steam entering on the right of A, while that on the left escapes to the atmosphere. The total pressure on the steam-piston equals the steam pressure per square inch above the atmosphere multiplied by the area of the piston, and that on the water end equals the water pressure multiplied by the area of the water-piston. These two total pressures are almost the same, the total steam pressure being greater by the amount of friction in the pump. Therefore, if we make the diameter of piston C sufficiently larger than that of the piston D, the water pressure

* See page 187 for full description of this action.

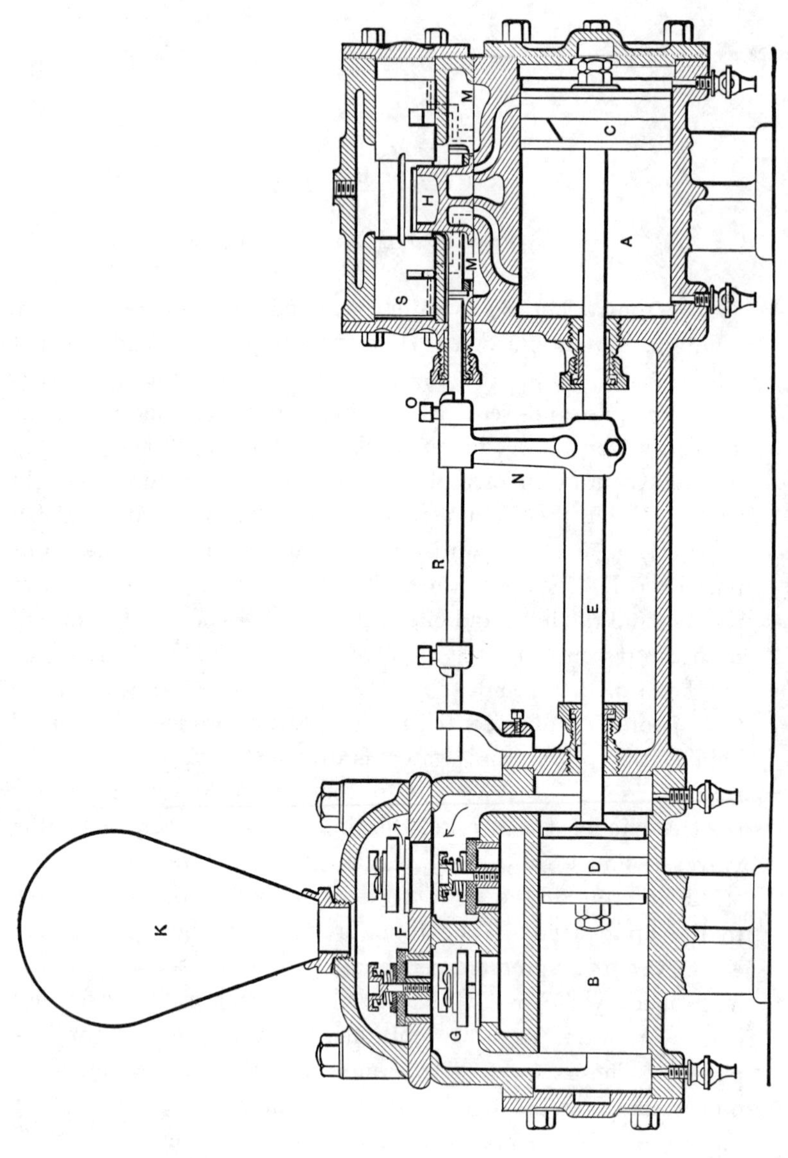

FIG. 79.—Single Cylinder Pump.

FIG. 80.—Warren Duplex Pump.

(*To face page* 73.)

will be so much higher than that of the steam that the discharge will overcome the friction of the pipes and valves and enter the boiler against the pressure of the steam which drives the pump.

The **air-chamber** K is placed on the pump to equalize the pressure in the discharge-pipe, as, when the pump reverses, there is usually a change in the velocity of the water, which produces a variation in the pressure, and is apt to cause a shock to pipes and pumps. The air which is above the water acts as a cushion, either keeping up the supply to the main when the pump slows down, or receiving the excess pressure when there is an increase in the discharge.

At times two pumps are placed side by side, as shown in Fig. 80, forming what is known as a **duplex-pump.** In this case the **piston-rod** of one pump controls the movement of the steam-valve of the other.

The water-piston shown in Figs. 78 and 79 is so constructed that rings of hemp, flax, or canvas indurated with rubber, can be placed between two discs and forced against the cylinder walls, making a water-tight joint. This canvas, or hemp, is called **packing,** and the piston is said to be **inside-packed.** When this packing is not tight, water will leak from one side to the other. This leakage cannot be seen, and, to make sure

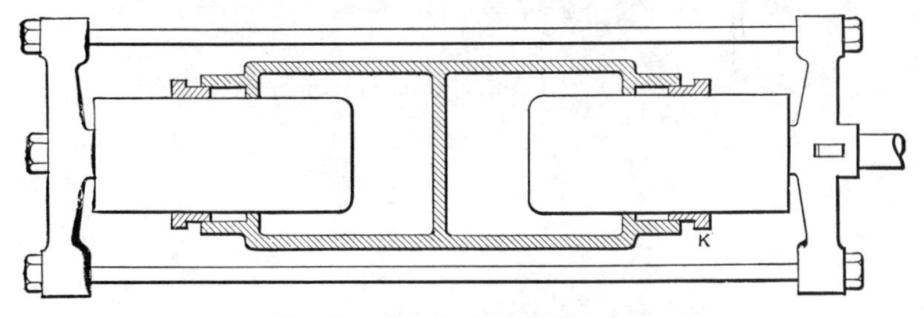

FIG. 81.—Outside Packed Plungers.

that there is none, the **outside-packed** plunger, as shown in Fig. 81, is used. This plunger is forced into the water contained in the pump-cylinder and thus drives it out. When the plunger is withdrawn there is a vacuum produced and water enters to fill the space. The only place water can leak around this plunger is at the **stuffing-box** K, which is packed in a manner to be described later. Should leakage occur, it is easily seen and the gland tightened to prevent it.

A pump is designated by the diameter of its steam-cylinder, the diameter of its water-cylinder, and the stroke. Thus, a $4'' \times 3'' \times 4\frac{1}{2}''$ pump is one in which the steam-cylinder is $4''$ diameter, the water-cylinder $3''$, and the stroke $4\frac{1}{2}''$.

Hot-well.—The pump may draw water from a vessel below the level of the pump, usually termed the **pump-well.** When, however, there is any hot water available, whether as condensed steam from the engine or from any other source, it is discharged into this well, which is then called a **hot-well.** Heat is thus saved, which increases the efficiency of the plant. As water is raised because of the vacuum produced in the water-cylinder of a pump, hot water cannot be lifted through

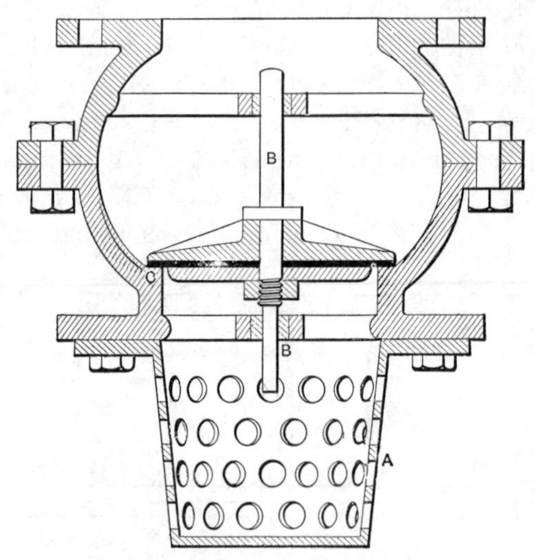

FIG. 82.—Foot-valve and Strainer.

any great height, as under the reduced pressure steam is given off, preventing the formation of a vacuum. In this case the hot-well should be placed as near the level of the pump as possible, if it cannot be placed above it.

A **foot-valve** is usually placed at the lower end of the **suction-pipe** of a pump. This valve, Fig. 82, is really a large check-valve to which a strainer A is often attached. The valve-disc is faced with leather and moves upward when the pump draws water. The spindles BB guide the disc. The leather is held between two plates and rests on the seat C, which is part of the casing.

Injector.—The general method of feeding locomotive boilers is by the **injector.** This is an apparatus in which steam

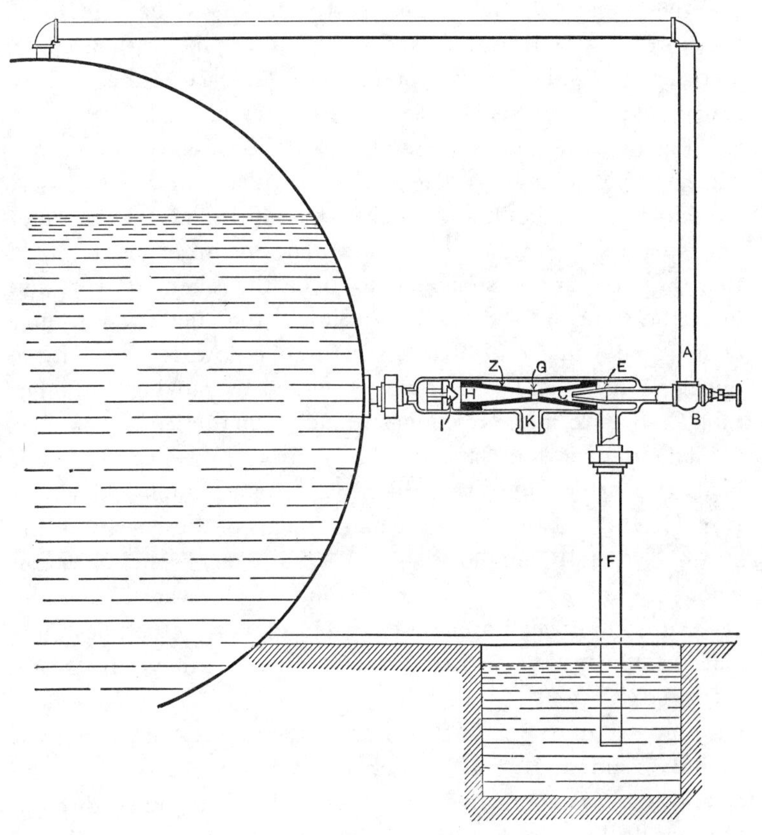

FIG. 83.—Diagram of Injector.

is allowed to obtain a high velocity, and, on being condensed by coming in contact with a stream of cold water, gives up part of its momentum, so that the mixture enters the boiler against the pressure of the steam. Fig. 83 shows the scheme of an injector. The steam enters at A, through the valve B, and in expanding from the boiler pressure to that at the nozzle C, part of the heat energy of the steam is changed to kinetic energy. The shape of the nozzle is such as to permit of the gradual increase of velocity, which is so great on leaving C that there is a vacuum formed in the chamber E and water is drawn up the suction-pipe F. The water condenses the steam, and the impact of the steam greatly increases the velocity of the water. When water is flowing in a pipe with a given velocity and under a given pressure, a decrease of the velocity is accompanied by such an increase of pressure that the sum of the kinetic and potential energies of the water remains constant. This assumes no losses from friction or other causes. In the injector the high velocity at G of the mixture of water and steam is so reduced by the enlargement of section at H that the pressure is sufficient to open the check-valve I and force the water into the boiler. Should too much water enter at F, some will be discharged at the overflow K through the openings at G. This may be prevented by moving the inner tube Z so as to make opening around C smaller.

The injector is of special value as a boiler-feeder because it heats the water and is very efficient. As a pumping device it is inefficient, because such a small amount of the energy of the steam is usefully employed. This, however, is no objection when we can use the heat remaining in the water, as is the case with boiler-feed apparatus. Injectors are sometimes difficult to adjust, but with a little experience they are easily managed.

Fig. 84 shows the **Little Giant Injector.** Water enters at A and steam at B. The discharge takes place at C. The steam supply is controlled by a globe valve, and enters through the nozzle D. In the **combining-tube** E it mixes with the water which has been lifted because of the vacuum produced. The mixture then enters the discharge-tube F which so increases in cross-section that the velocity of the water is quite low when it reaches the check-valve in the discharge-pipe. The pressure at G is usually below that of the atmosphere, but if too much water is drawn into the injector the pressure in G rises and the water under this higher pressure opens the **over-flow-valve** H, discharging through the **overflow** K. To cor-

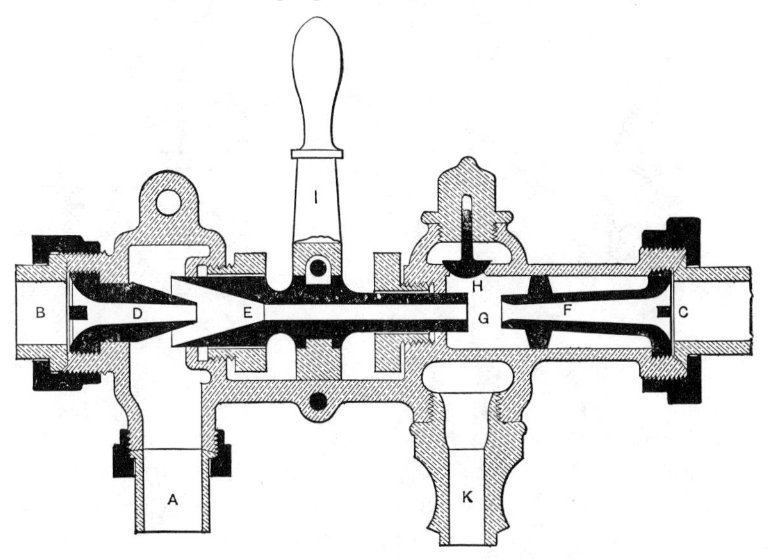

FIG. 84.—Little Giant Injector.

rect this, the combining-tube is moved to the left by the handle I, thus partly closing the opening between the nozzle D and the combining-tube, and so checking the supply of water.

To start the injector the water-valve is opened, and then the steam is turned on through the steam-valve. This raises the water, which mixes with the steam in the combining-tube, and discharges through the overflow until it attains such a velocity that it will raise the check and pass into the boiler. If the combining-tube is not correctly placed water will continue to issue from the overflow, and the tube should be moved by the handle I until there is no discharge. To stop the injector the steam is shut off and the water-valve closed. This injector will not raise its supply of water through a very great height, and where a supply has to be lifted an auxiliary apparatus is employed.

Sellers' Improved Self-acting Injector.—Another form of injector is shown in Fig. 85 and Fig. 86. It belongs to the

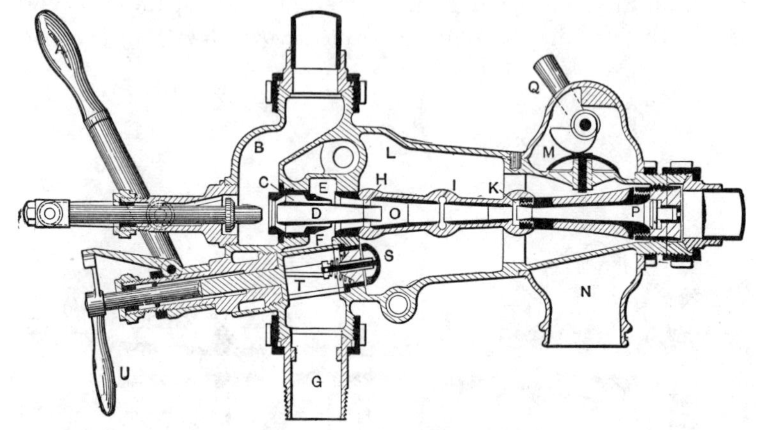

FIG. 85.—Sellers' Injector—Section.

automatic restarting type. Should the water-supply to this injector be broken from any cause, the injector would restart on the supply being re-established, and, as the steam pressure changes, the quantity of water is automatically regulated so that the discharge continues. This is not the case in the simple injector.

To operate this injector, the steam-supply valve is opened and the handle A is pulled to the left a short distance, which admits steam from the pipe B through the small openings C. This steam, in passing between the nozzles D and E, creates a vacuum in the space F and raises the water from the supply-tank through G. This water escapes through the openings H, I, and K into the space L and thence through the overflow-valve M to N. When water appears at the overflow the handle A is drawn further to the left, allowing steam to enter the central portion of the nozzle D, and this, coming in contact with the ring of water, combines with it in the combining-tube O and finally enters the boiler, passing the check-valve P. Should the supply of water be discontinued from any cause, the small jet through the openings C will continue to produce a vacuum in F, so that water will be immediately drawn up if the supply is re-established. The overflow-valve M can be held down by the cam and lever Q. This is done

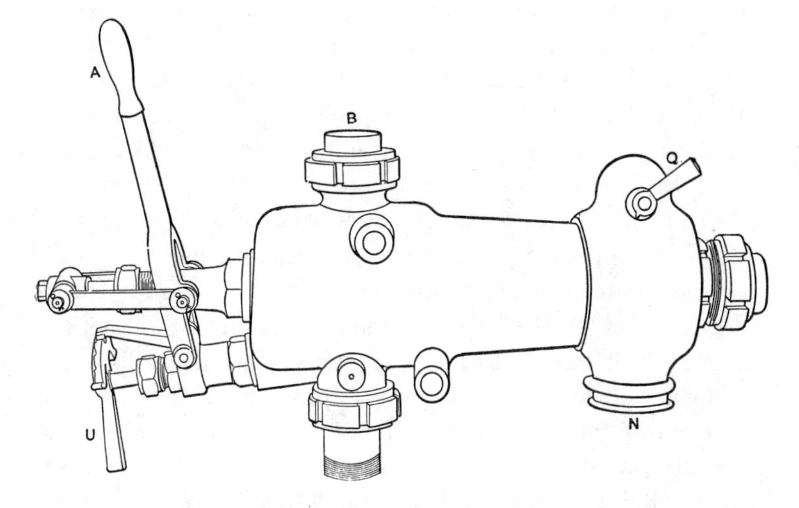

FIG. 86.—Sellers' Injector.

when it is desired to heat the supply or to clean the strainer on the supply-pipe. The valve being closed steam is admitted to the injector and fills the space F and passes down the suction-pipe.

When the injector is in operation, and the steam pressure rises, a partial vacuum is produced in the space L which draws water through the valve S. This water enters the small holes H, I, and K, and is forced into the boiler. With a reduction of the steam pressure the vacuum in F falls and less water is raised to unite with the steam in the combining-tube. The valve T, operated by the handle U, serves to regulate the quantity of water fed to the boiler.

Feed-water Heaters.—To save some of the heat which would otherwise be wasted in the exhaust-steam from engines, and to partially purify the feed-water, **feed-water heaters** are used in modern boiler plants. They are of two types, the **open heater** and the **closed.** In the former steam comes in contact with the water, and providing there is sufficient steam it raises the temperature of the water to 212° Fahrenheit, the excess steam passing to the atmosphere. In the closed heater the water to be heated is not open to the atmosphere, and the temperature can be raised to boiler temperature if desired. In an open heater the heat of the exhaust-steam is partially saved and the water enters the boiler purer, because this heating drives off any carbonic acid gas and causes some of the dissolved carbonates to be precipitated. If the water is then allowed to move slowly, the sediment will collect and the clear water can be drawn off. There are certain impurities which do not separate at this low temperature, and for them some form of heater must be used in which the temperature will reach the point at which these salts are precipitated. A heater in which the steam and water do not mix is apt to have scale form on the metallic surfaces, and hence become inefficient. **The Cochrane feed-water heater and purifier** is a type of open heater. It is shown in Figs. 87 and 88. The cast-iron sides A are strongly ribbed and bolted together, the joints being **rust-joints,** calked on the inside. At the proper points stay-rods hold opposite sides together. Water enters at the top through the regulating-valve B, and is distributed over the trays C, from which it finally drops into the settling-chamber. The water is broken up as it runs over the trays, or from one tray to the next, so that it comes in close contact with the steam which fills the upper portion of the shell. On these trays the deposits are precipitated. The mineral matter, or other impurities which fall from the trays, is filtered out by the filtering-bed of crushed coke X at the bottom of the shell. Finally, the water is drawn from the outlet D after passing through the perforated plate E. The casing contains a large, hollow ball-float F, which actuates the regulating-valve B, opening it when the level of the water falls, and closing when the water reaches the proper height. Fig. 87 shows the arm and shaft of the

FIG. 87.—Cochrane Feed-water Heater.

(*To face page* 80.)

bell-crank lever to which the float is attached, and the rod leading to the regulating-valve. This rod contains a turnbuckle G, Fig. 88, by which the height of the water can be adjusted,

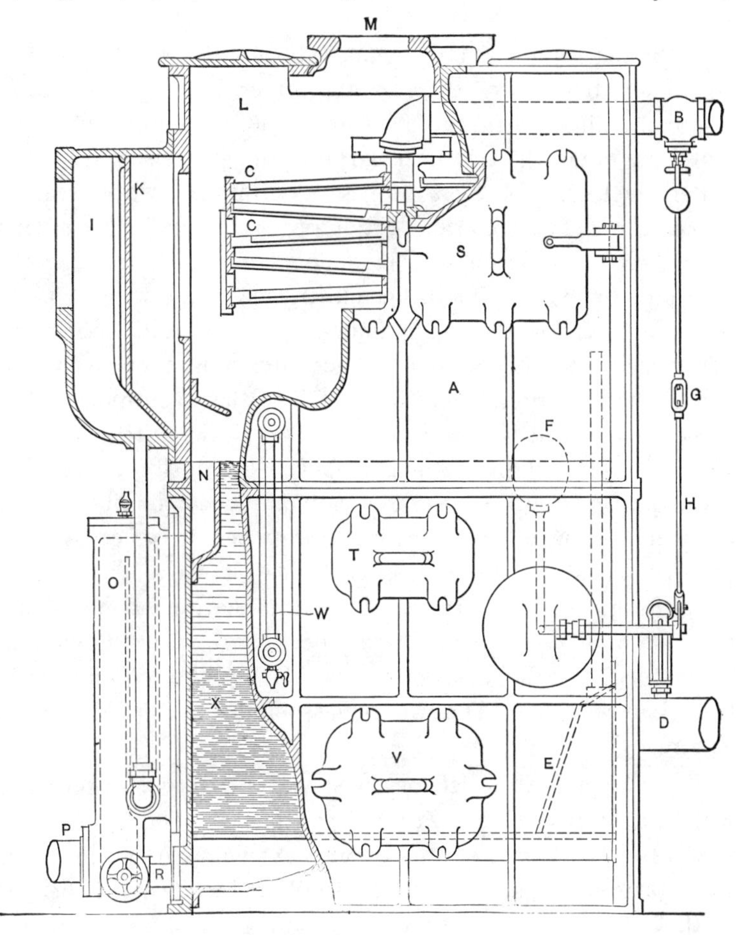

FIG. 88.—Section of Heater.

as the changing of the length of the rod H necessitates a new position of the float when the valve is closed.

Steam enters at the inlet I, and striking against the plate K, containing ribs, deposits its oil, and on passing around the sides of K enters the steam-chamber L. It has a free path to the steam-outlet M, but should there be much cold water on the trays the steam is drawn in that direction because of the steam condensation and consequent reduction in pressure.

N is the skimmer, an opening above the high-water level by which the oil or other floating impurities can be skimmed off by holding down the float-arm and thus allowing more water to enter than is necessary for the pump supply. The water-level rising, the excess passes off into the **water-seal** O, carrying with it the matter floating on its surface. The oil and water, separating from the incoming steam, fall through the drip-pipe to the water-seal O, and thence to the waste-pipe P. This water-seal is a box containing a partition, the water being introduced near the bottom. The water must then rise to the top of the partition before it can fall to the waste-pipe. The object of the seal is to prevent air being drawn into the heater when there is a slight vacuum produced during low steam-supply. At the lower part of the shell is placed the blow-off R, by which the water may be drawn off when it is necessary to clean the heater. In front of each nest of trays are the tray-doors S, and lower on the shell are the charging-doors T and the manhole door V. These doors are so arranged that when necessary the heater can be quickly cleaned and again placed in service. The water-glass shows the height of the water.

Fig. 89 shows the **Hoppes feed-water heater** and **purifier,** which is formed of a series of trays in a shell, from each of which the water passes into the one next below in a thin sheet. Live steam is introduced into the shell, and the water is heated so that the salts are precipitated and deposited on the top and bottom surface of the trays, while the sediment in the water has a chance to settle. The door swings out of the way to permit the removal of the trays.

The **Wainwright feed-water heater** (closed heater) is built with a number of tubes, through which the feed passes while the steam passes around the outside, as shown in Fig. 90.

Fig. 89.—Hoppe's Heater.

(*To face page* 82.)

Water enters at the bottom at A and leaves from the top at B, passing through the corrugated copper tubes which are surrounded by exhaust steam. The steam enters at C and leaves at D. Blow-off pipes are connected at E and F, and a safety-valve at G, to relieve the pressure if both inlet and outlet valves are closed.

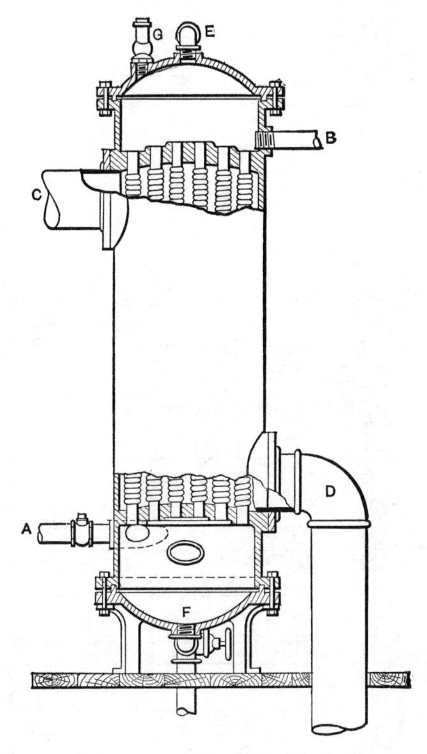

FIG. 90.—Wainwright Heater.

Green Economizer.—To save some of the heat which is carried away by the gases escaping from the boiler various forms of economizers have been proposed, differing mainly in details. They consist of a series of pipes placed between the boiler and the stack, through which the feed-water passes. The temperature of the water is raised, and, as the water is under more

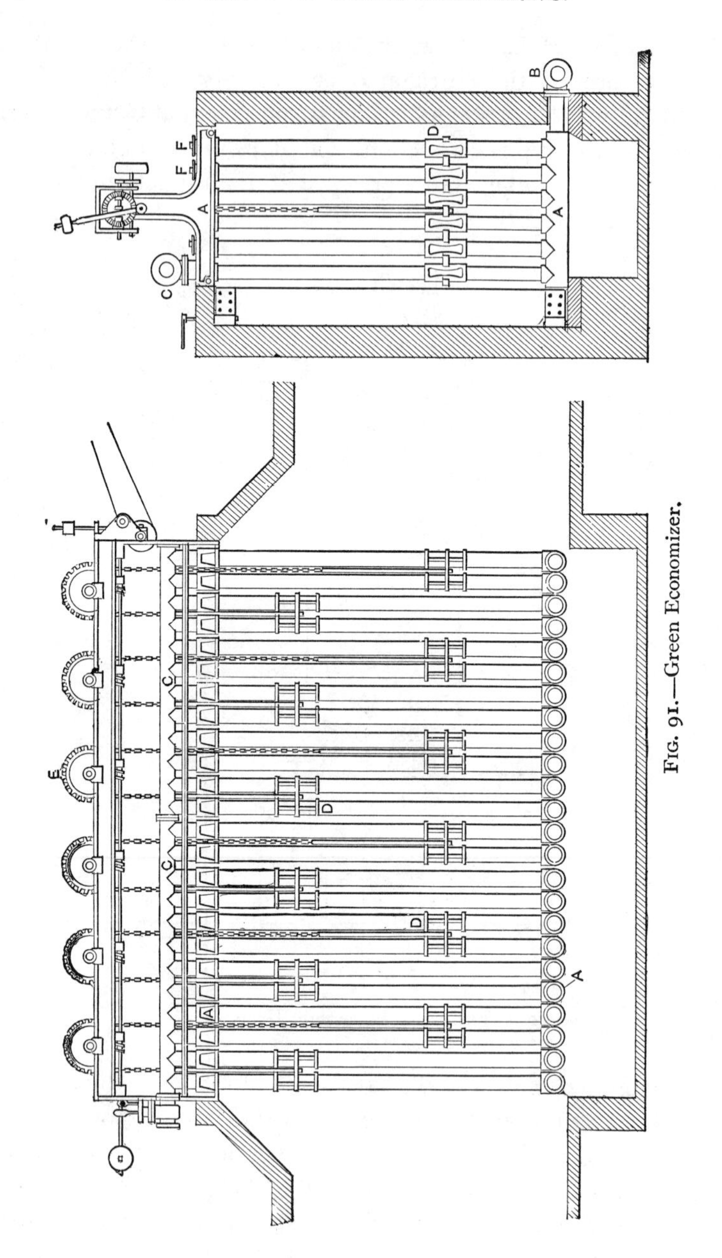

FIG. 91.—Green Economizer.

than atmospheric pressure, it is possible to have the feed temperature above 212° Fahrenheit. The economizer should not be built too large, as it would then destroy the chimney draft by cooling the gases too much, and hence render it necessary to use some form of blower.

The Green Economizer consists of sections of pipe, each section being fastened on two castings A, called manifolds, shown in Fig. 91. These manifolds serve the same purpose as the headers in a water-tube boiler; that is, they distribute the water at the lower end of the pipes or tubes and collect it at the top. The manifolds are all joined by two longitudinal manifolds BC, one at the top and another at the bottom. Water enters the various sections from the pump, and rising through the tubes, which are about four inches in diameter and nine feet long, is heated and conducted to the boiler through the collecting-manifold C at the top.

The tubes are furnished with **scrapers** D, which are raised and lowered by the chains which pass over the **chain-wheel** E at the top. These chain-wheels are driven by gearing actuated by some auxiliary apparatus. That the tubes may be cleaned of sediment and scale, which will collect from water containing salts in solution, the top and side manifolds contain caps F. The economizer is placed in the flue leading to the chimney, and when necessary to repair it, the gases are conducted to the stack through an auxiliary flue.

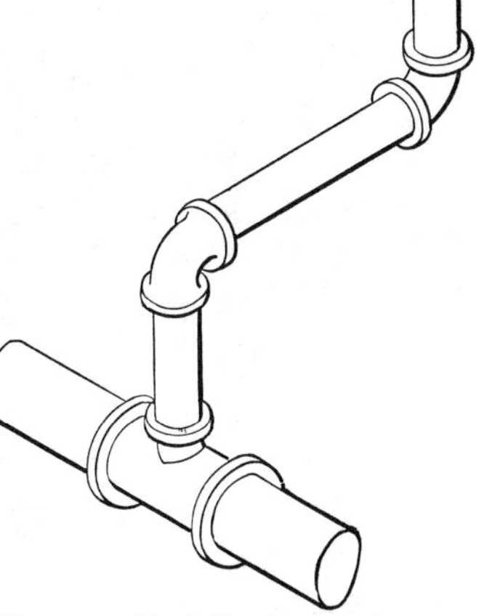

FIG. 92.—Feed-line Connections.

Feed and Blow-off Lines.—In large plants the feed for all of the boilers is taken from a main feed-line, which extends the length of the boiler-house. This line is connected to each boiler by a branch, and the connection should be made as shown diagrammatically in Fig. 92, so that, when hot water is used there will be no danger of breaking the pipe-fittings from the expansion of the line. A pump supplies the line with water, which may have been drawn from an open heater or forced through a closed heater or economizer. It is a good plan to have two pumps attached to the feed-line, that one might be overhauled without crippling the service. In all installations duplicate machines, or parts, are advisable, to be used in case of emergency. Injectors are sometimes supplied in addition to feed-pumps, and these are often connected with a separate supply-pipe.

As the steam-piston of a pump is so much larger than the pump-plunger, the water pressure becomes very high when all the feed-valves on the branches are shut off. A safety-valve or a **pump regulator** should be used as a safeguard against the breaking of the feed-line. A pump regulator is so constructed that when the pressure in the feed-line becomes high the pump is automatically shut off.

The **blow-off pipes** from each boiler are connected to a main blow-off, and this is led to a large tank, called the **blow-off tank.** This tank is supplied with a drain and a large vent to discharge the steam before the water enters the drain. A blow-off tank is always advisable to reduce the velocity of the issuing water. If the blow-off discharged directly into a sewer the steam produced would cause much annoyance, and the scouring action of the stream of water would probably ruin the sewer. This practice is generally forbidden.

Draft.—To burn fuel, sufficient air must be brought in contact with it to unite with the carbon and hydrogen. The air may be drawn through, in which case it is said to be **induced,** or it may be forced through the fuel, giving the system of **forced draft.**

Natural Draft.—As air is heated it expands, and so becomes lighter. If this heated air is allowed to issue into a colder atmosphere it will immediately rise, as it is lighter than the surrounding air. If a chimney is made so that the gases will not lose their heat, and if this chimney be filled with heated gases, their weight is so much less than that of the same height of cool air outside that there is a strong tendency for the heated gases to ascend. This produces a partial vacuum in the furnace and so draws the air through the fire.

Natural draft, although simple in theory, is costly, as the gases carry off from twenty to thirty per cent of the heat in the coal. If an attempt is made to save this heat by economizers the draft is partially destroyed, and hence schemes have been proposed to abstract all of this heat by an economizer and use a fan at the base of a very short stack to draw air through the fire and the gases through the economizer. This method of induced draft has been installed in several large plants, but its real saving is questionable.

The economizer must be kept clean both internally and externally, and the engines and belts driving the fans must be cared for. In place of a large stack a small inexpensive one is used, and in place of the gases entering the stack at from 400° Fahrenheit to 600° Fahrenheit, they enter the flue at about 200° Fahrenheit. These are both items of economy. Of course an economizer, an engine, and a fan are required, and partly offset the cost of the stack, and the cost of repairs, attendance, and steam is considerable.

Fig. 93 shows the construction of a **steel stack.** These stacks are quite extensively used because of their cheapness and ease of construction. The stack is supported on a **foundation** of concrete or stone masonry, being fastened to it by heavy bolts about $2\frac{1}{2}''$ in diameter. These bolts are attached to brackets riveted to the lower ring. The rings are all made of steel plates riveted together with lap-joints. The **lining,** or **core,** of the stack may be carried by the lower part of the stack, or may be supported by the foundation. It is made of different thicknesses of brick masonry, which may be faced with fire-brick, although there are instances in which red bricks have been used throughout. A ladder is usually riveted to the stack, so that the top may be reached when necessary.

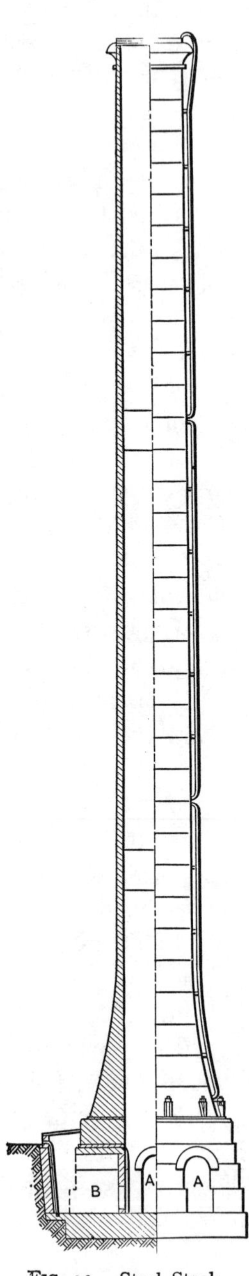

FIG. 93.—Steel Stack.

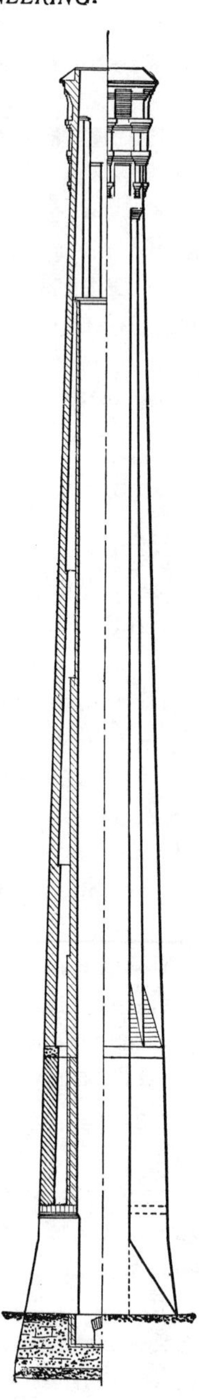

FIG. 94.—Brick Stack.

A terminal band is usually added at the top for a finish. The flue enters a large opening A at the bottom, while at B is an opening by which the chimney and flues may be cleaned. Small stacks are sometimes built without linings, but this is not advisable. The stack shown is 170 feet high, 10 feet inside diameter, and intended to be used for 2500 horse-power. The steel plates vary from $\frac{3}{8}''$ to $\frac{1}{4}''$ in thickness, the courses being 5 feet wide. The rivets in the horizontal seams are $\frac{3}{4}''$ in diameter and of $3\frac{1}{2}''$ pitch. This seam is double-riveted while the vertical seam is single-riveted. A vertical section of a **brick stack** is shown in Fig. 94. This stack has a core separated by an air-space from the walls of the chimney. Both the stack and the core should be protected from the weather at the top. This stack is 140 feet high, 8 feet inside diameter, and is intended for 2500 horse-power. The outer wall varies from 26 to 9 inches, and the core from 13 to 4 inches in thickness.

Induced Draft.—Another system of induced draft is that used on the locomotive. The steam discharged from the cylinder is delivered through a nozzle, so shaped that the velocity of steam at exit is high. This discharge occurs below the base of the stack, and draws with it gas from the smoke-box, producing a partial vacuum in the tubes and fire-box, so that air is drawn through the fire. In this way so much air is supplied that an intense fire is produced. So strong is this induced draft that with soft coal (bituminous) the rate of combustion sometimes reaches 200 pounds, while with natural draft from 20 to 30 pounds is considered excellent. With hard coals this rate is much lower, being from 10 to 15 pounds per square foot of grate per hour with natural draft, while in locomotives it is from 75 to 125 pounds. Fig. 95 shows a **blast-pipe.** AB are the exhaust-passages from the two cylinders, which enter the two boxes CD, forming the exhaust-nozzle. The nozzle terminates in two circular openings, into which the thimbles E are placed, making the velocity of the exhaust-steam greater, and thus increasing the draft.

On tug boats and in steam fire engines the exhaust is discharged in the stack to aid the draft. A steam jet is sometimes discharged in the stack to be used when the engine is at rest or when the natural draft is not sufficient.

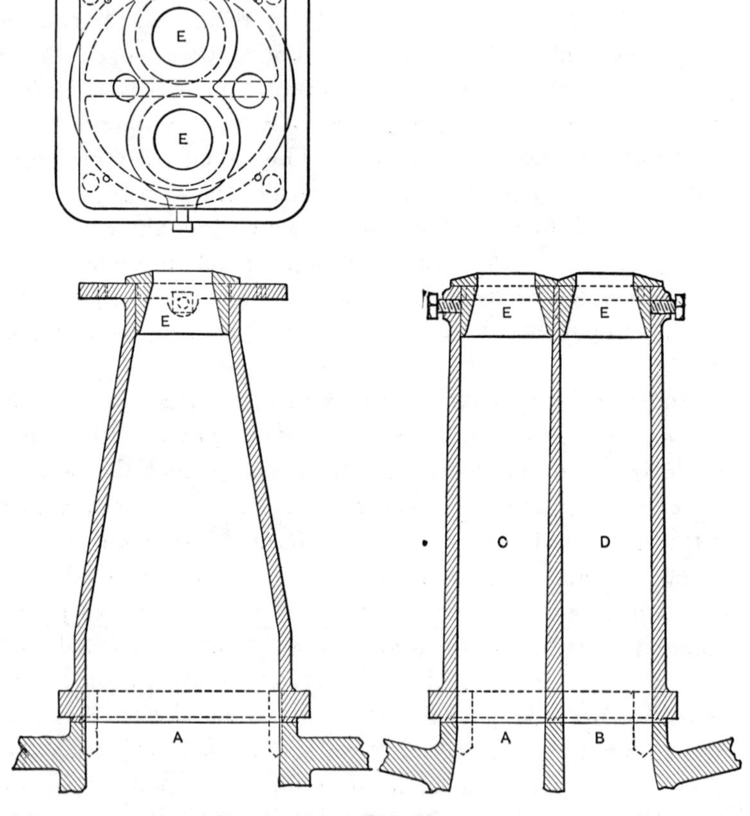

FIG. 95.—Blast Pipe (Locomotive).

Forced Draft.—With marine boilers forced-draft systems are used. In one system the air is forced into the ash pit, which is closed, and in another, the air is forced into the fire room or stoke hold. These are called respectively the **closed ash pit** and the **closed stoke-hold** system of forced draft.

In the first system the air is drawn from the outside or from the stoke-hold by a fan-blower, and then, as in the **Howden system,** Fig. 96, it is forced around a series of tubes which

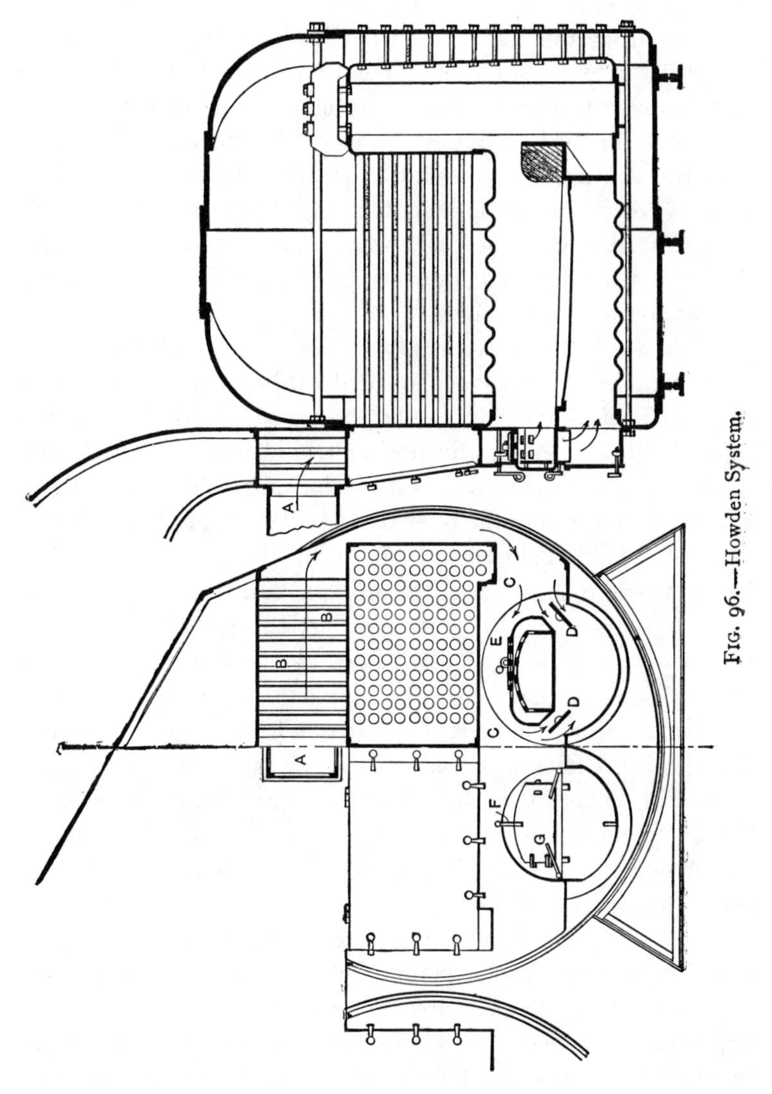

FIG. 96.—Howden System.

are surrounded by the hot gases leaving the boiler. The air enters the ash-pit and passes through the grates. Some of the heated air is allowed to enter above the coal, and thus to burn any unconsumed gases.

Fig. 96 shows this system applied to a Scotch marine boiler. The air from the fan enters from the pipe A at the center of the boiler and passes around the tubes B to the outside, and thence down to the reservoir C around the fire-doors. From this reservoir the heated gas can pass the **butterfly-valves** DD into the ash-pit, or through the **gridiron-valve** E to the fire-box. The valve E is formed like a gridiron, with bars which cover slots in the surface over which it moves. The butterfly-valves are similar to dampers. Handles FG in front of the casing control these valves.

When it becomes necessary to fire more coal the draft is shut off and the fire-door opened. Should the draft not be closed off, the flame from the fire-box would be driven into the stoke-hold on opening the fire-door, and so endanger the life of the stoker. To prevent this, a locking device is used, so arranged that before the door can be opened the draft must be shut off. With a closed stoke-hold this danger does not exist, as the whole room is under the higher pressure, air being forced in by blowers. All the openings into this room are made by double doors forming "**air-locks,**" so that entrance can be effected without permitting the confined air to escape. It is claimed that this method makes a cooler fire-room, but should the air for the closed ash-pit system be taken from the room, there will be just as much fresh air brought into the stoke-hold. An objectionable feature of the closed stoke-hold system is that while firing, there is a great inrush of cold air, which is apt to loosen the tube-joints.

In the Howden System the heated air not only saves some of the heat which would go off with the chimney gas, but also produces a more intense fire. In either of these systems, as the draft is produced by the blower, it is not necessary to depend on the heated gases for draft. The object of forced draft is to increase the rate of combustion, thus producing more steam from the same boiler. This is necessary on war and merchant vessels where high speed is desired, because the space occupied by the additional boilers, to give the necessary steam, cannot be allotted to the engineering department.

Damper Regulators.—The fireman watches the water-column on the boiler and adjusts the feed-valve so that the level is nearly constant. To keep the pressure of the steam constant he must move the damper, opening it when the pressure falls, or closing it when it rises. If more steam is required than a boiler is producing, the pressure will gradually fall, and, to prevent this, coal must be burned at a higher rate of combustion. This is accomplished by opening the damper and thus allowing more air to be drawn through the fire, consuming more coal. Should the demand for steam decrease, the steam pressure would rise and the damper would be closed to accommodate the new conditions. To do this automatically, and to keep the pressure more nearly uniform, the damper regulator has been invented.

The **Patterson damper regulator** is shown in Fig. 97. It consists of a cylinder A, with the piston B, the piston-rod being connected with the damper of a boiler, or with a series of dampers, by the wire rope C. Water enters A from the pipe D through the regulating-valve E and pipe F, raising the piston and so closing the damper. This admission of water is controlled by the valve E, which is pushed upward by the spring G when the arm H is raised. In the position shown in Fig. 97 the valve is pushed down, allowing the water in A to be discharged through the overflow I by the pressure exerted by the weights J. When the valve is raised by the spring the upper disc of this balanced valve closes off the overflow and allows the water from F to enter D through the center of the valve. The arm H is attached to the clamp K, which slips over the lever L and is fastened by a thumb-screw. This lever, having a fulcrum at M, is raised by the piston in the cylinder N. This piston rests on a **flexible diaphragm,** which is bolted between flanges, thus making a movable joint not liable to leak or corrode. The pipe O, which is filled with water, is connected to the steam-line, so that the water is under the pressure of the steam, and this acting on the bottom of the diaphragm tends to lift the lever L against the weights P and Q. When the steam pressure rises sufficiently the lever is raised until it strikes the upper stop at R, allowing the valve to rise. This admits water under pressure into A, and thus moves B upward, closing the damper. As soon as the press-

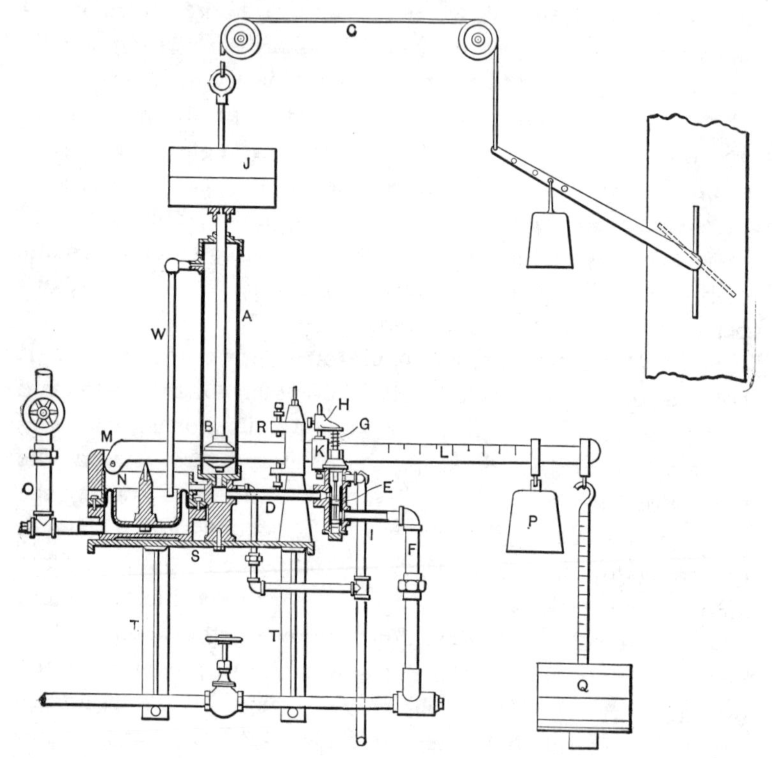

FIG. 97.—Patterson Damper Regulator.

ure falls the lever drops, and this forces the valve downward, discharging the water from A and opening the damper. The regulator is carried on a shelf S and the brackets T. The fitting below F contains a strainer to prevent the valve E from becoming clogged, and W is an overflow to remove any water that may leak past the piston. This water is discharged through a pipe into a main overflow I. The pressure of the water in pipe D does not affect the setting of the regulator, as it acts on the two discs of the balanced valve E and produces no pressure on the lever. Around the valve is a bushing, forming the ports; this is easily renewed, when worn, by removing the caps from the top and bottom of the valve-case.

Conveyers and Bins.—The method of handling and storing coal has undergone great changes in the last ten years. In many stations the coal and ashes are not touched from the time the coal leaves the cars to the time the ashes are dumped into the railway car for removal. This has been rendered possible by the invention of the conveyer and of the automatic-stoker.

The coal is brought to the fire-room in carts or cars and dumped into a bin below the driveway. From this point it falls into the buckets of a conveyer, which are drawn by an endless chain to a high level, where they discharge into bins over the boiler. Coal which is not uniform in size is usually passed through a crusher before passing into the conveyer. The coal then falls through spouts into the hoppers or magazines of the automatic-stokers. The ashes from the stokers drop into a hopper and are carried by a conveyer at the proper time to a convenient storage-place, from which they can be discharged into carts or cars. There are many modifications of this system to meet some special requirements, or to conform with the ideas of some engineer or designer.

Fig. 98 shows a plant installed by the Link Belt Engineering Co., in which a boiler-house of two stories is supplied with coal through spouts from the bins above the boilers. The same conveying-buckets which carry coal are used for ashes, which are discharged when necessary from hoppers beneath the boilers. The ashes are dumped into their storage bin by moving the tripping apparatus, which causes the bucket to discharge at any desired point.

In using coal an accurate account should be kept of the amount used, so that the cost of running the plant can be ascer-

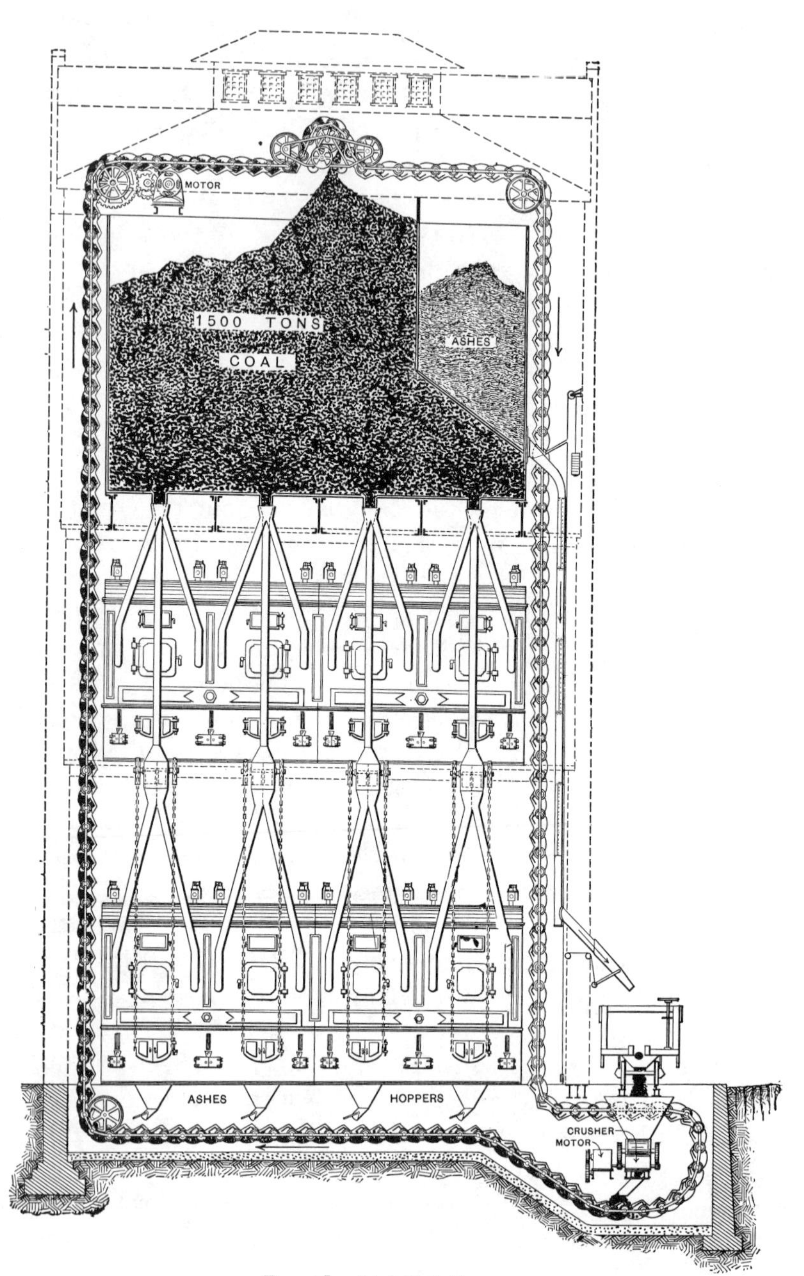

FIG. 98 —Link Belt Conveyers.

tained, and for that purpose automatic scales have been designed which work in conjunction with the coal-handling apparatus.

Steam-line.—The various boilers in a plant are usually connected to a large steam-line, which conveys the steam to the engine-room where it is distributed to the various engines. This line should be of ample size and carefully erected, provision being made for **expansion** and **draining,** two of the most important considerations in designing a steam-line. When a pipe-line is carrying steam at one hundred pounds pressure the temperature is such that the expansion in fifty feet is about one inch, and the connections must be so made that there is no danger of breaking fittings by pulling them out of line. To make this clear, consider the exaggerated case in Fig. 99,

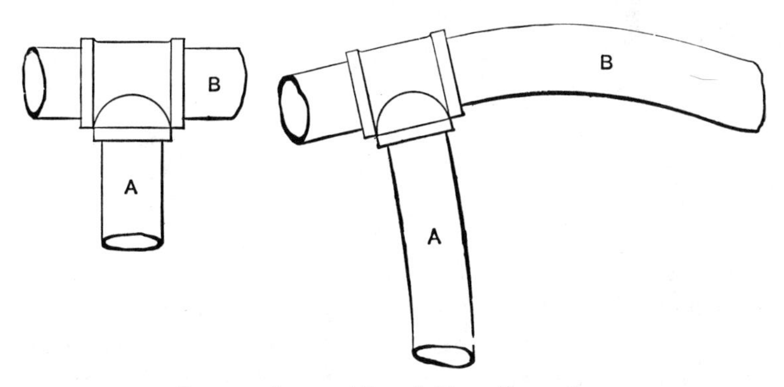

Fig. 99.—Incorrect Branch Steam Connection.

which shows a branch A taken from the main steam-line B. In the right-hand diagram the main line has moved to the left of its original position, and the cast-iron fitting which unites the branch to the main tends to keep them at right angles. When heated the pipes are bent as shown and this often breaks the "tee." To avoid this the joint should be made as shown in Fig. 100, with elbows AB, the nipple C, and the pipe D. The nipple turns in the thread of the elbow and allows the expansion to occur without breaking any of the fittings. In connecting the separate boilers to the steam-line this method is used, although bends with a stop-valve at the top, as shown in Fig. 55, are often used. The threads in the flanges at the bottom of the bends will give, and the bend itself will permit of some displacement.

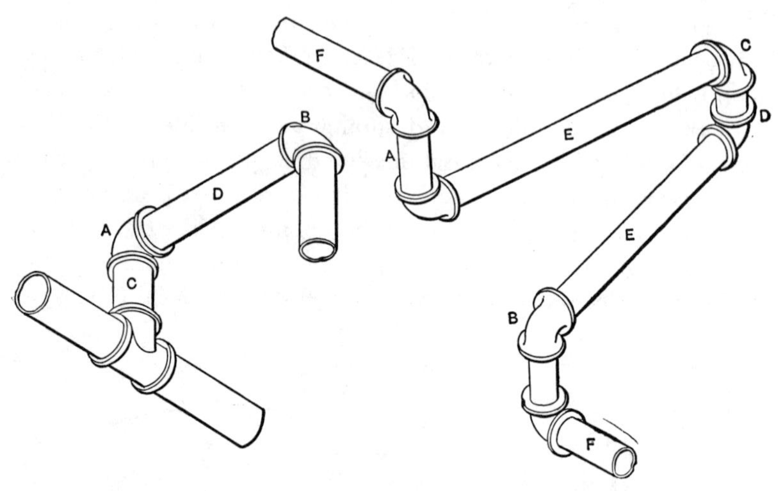

FIG. 100.—Correct Branch Steam Connection.

FIG. 101.—Swing Ells.

Expansion-joints.—When a point on the line is to remain stationary it has to be **anchored** by means of straps to some fixed point, such as a wall or girder, and if two such points occur on the line, some form of an expansion-joint must be used to take up the expansion between them. When the line is long, it is advisable to divide the expansion between several parts.

A simple form of expansion-joint is shown in Fig. 101. Here the joints at A and B or C and D move and allow the pipes EE to open out as the pipes FF contract and separate. This form of joint is cheap and effective when properly constructed. Another form is the **expansion-bend,** as shown in Fig. 102. It is made of copper or iron, and by bending allows the movement of the pipes to which it is attached. The **corrugated copper expansion-joint** is shown in Fig. 103. This allows for the expansion in a pipe-line by the compression of the corrugations. A form shown in Fig. 104 is used when none of the preceding are applicable. The body A is of

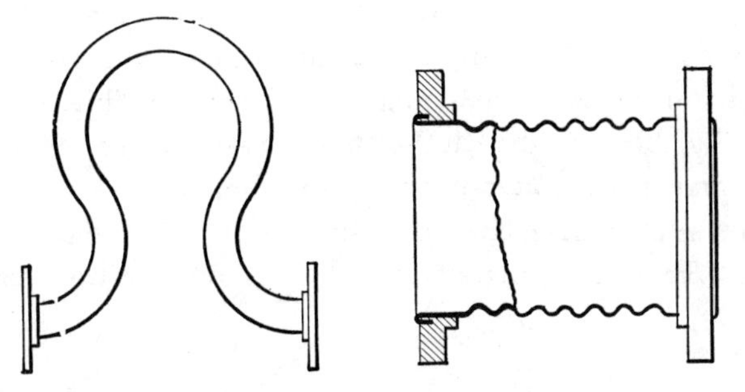

FIG. 102.—Expansion-bend. FIG. 103.—Corrugated Expansion-joint.

cast iron with a brass sliding-sleeve B. The sleeve slides over the body A as the pipes change length. To keep the steam from leaking around the pipe the space C which is left between the two pipes is filled with some form of hemp or pat-

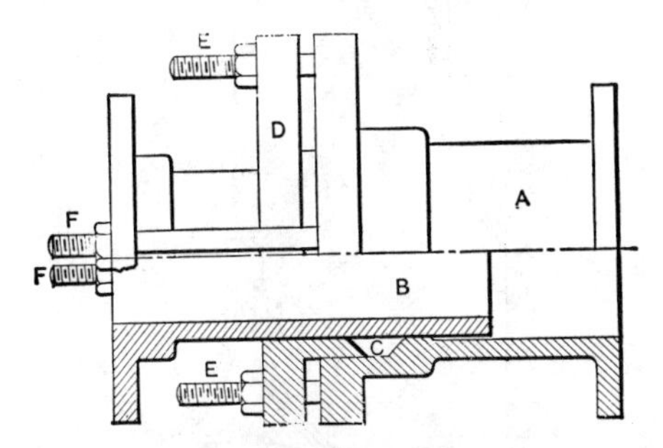

FIG. 104.—Packed Expansion-joint.

ent packing, pressed tightly against the pipe by forcing in the gland D. The gland is moved by the bolts and nuts E. To prevent this expansion-joint from separating if the anchors fail to hold, or where two joints occur without an anchor between, safety-bolts FF are provided, which limit the amount that the joint can open out.

Location of Valves.—Steam-valves should always be placed on the highest point of the line, so that no water can collect on either side of them. If placed at a lower point the upper side should be drained, so that no water can remain there. The pressure side of the valve should always be the lower side in case of a globe valve, so that when closed the stuffing-box can be easily repacked, which could not be done with steam around the spindle. When possible the valve should also be placed with the stem horizontal, as this to some extent prevents a water-pocket being formed.

Pipe-supports.—The pipe-line is usually supported by **hangers,** Fig. 105, carried on brackets on the wall or from

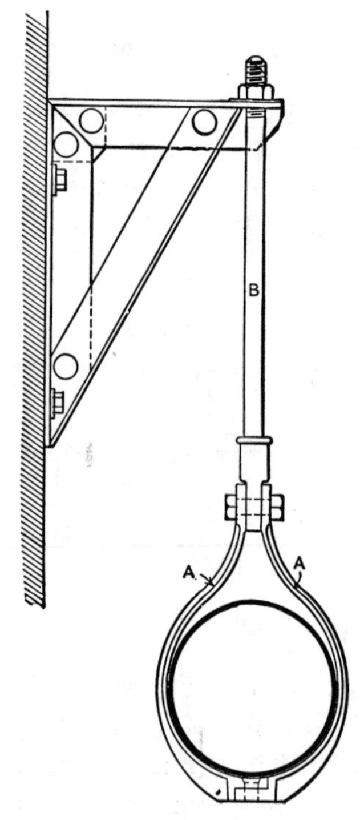

FIG. 105.—Pipe-hanger.

girders, and in some cases by columns extending from the floor. These hangers A should be so made that the pipe can move freely. To accomplish this the suspension-rods B are long, or, when columns are used, the pipe is so held that it can easily slide or move on rollers.

Drips and Traps.—The steam-line should be erected with a gradual inclination downward in the direction in which the steam is moving. The water tends of itself to move in this direction and the steam aids its motion. If the inclination is in the opposite direction the flow of steam opposes that of the water, and the water is apt to "**back up**" in the pipe. When the velocity of the steam changes, this water is lifted and carried with the steam, frequently causing much damage. The low points of the steam-main are drained by drip-pipes which conduct the water to traps, from which it is discharged to the hot-well. Steam outlets should always be taken from the top or bottom of a pipe on account of expansion, as shown in Fig. 100, and, if possible, the top should be used, as there is then no danger of receiving, in the branch, the water which flows along the bottom of the main pipe. The end of the line should always be dripped.

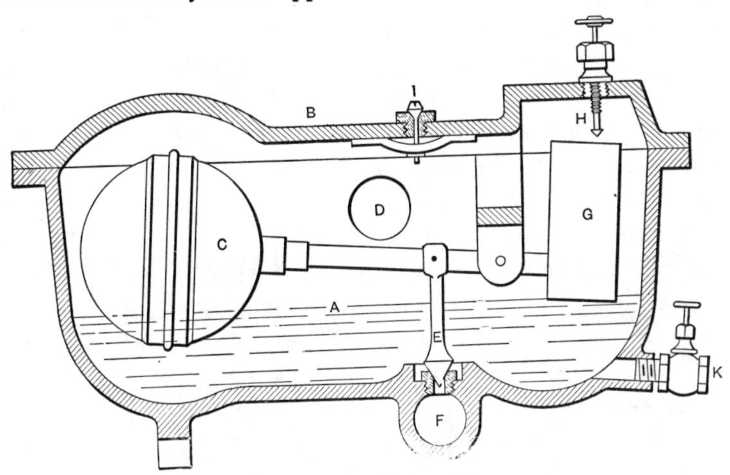

FIG. 106.—McDaniel Steam Trap.

A **steam trap** is an apparatus by which water is allowed to drain from a steam system without the loss of steam. The trap shown in Fig. 106 is one of the many forms which are found on the market. The **McDaniel trap** consists of an iron box A, with the cover B bolted to a flange. The box contains a **ball float** C, which rises as water enters through D. As the float rises, the **spindle valve,** or **needle valve,** E, which is attached to the arm of the float, is lifted from its seat, and the water is forced out through the discharge-pipe F. Before the water is all driven from the box, the float falls and closes the valve. The counter-weight G balances part of the weight of the float. To clean out the trap or to keep it open, the spindle H is supplied. By screwing this down the valve is raised and the entire trap is drained. When the trap is filled with air, or **"air-bound,"** no water can enter, and the air must be driven out of the trap by opening the discharge or the air-valve I on top of the trap. The blow-off valve K also serves to drain the trap.

Pipe-covering.—To prevent the excessive condensation of steam in a system, the piping is usually covered with some non-conducting material. These pipe-coverings are made of various substances, such as compressed cork, hair-felt, asbestos fibre mixed with magnesia, etc. These materials are molded in halves to fit the pipe or fitting, or they may be applied in the form of a plaster. To finish the exterior they are covered with canvas, or a smooth finish is made with "plaster of Paris." These coverings are of great value in preventing loss of heat by radiation, as well as in making the engine-room cooler. The amount of steam condensed by 1 square foot of exposed pipe, containing steam at 100 pounds pressure, is about 18 pounds per twenty-four hours, while, with a good covering, this may be reduced to 4 pounds.

Separators.—The steam-line is continued to the engine-room, and at this point a **steam separator** should be placed to remove any excess of moisture from the steam. Fig. 107 represents a **Cochrane separator.** It is formed of a cast-iron body A, with a reservoir B bolted below. To the two flanges C and D on the body are bolted the flanges of the steam-pipe. The body A contains a baffle-plate E, on which are the ribs

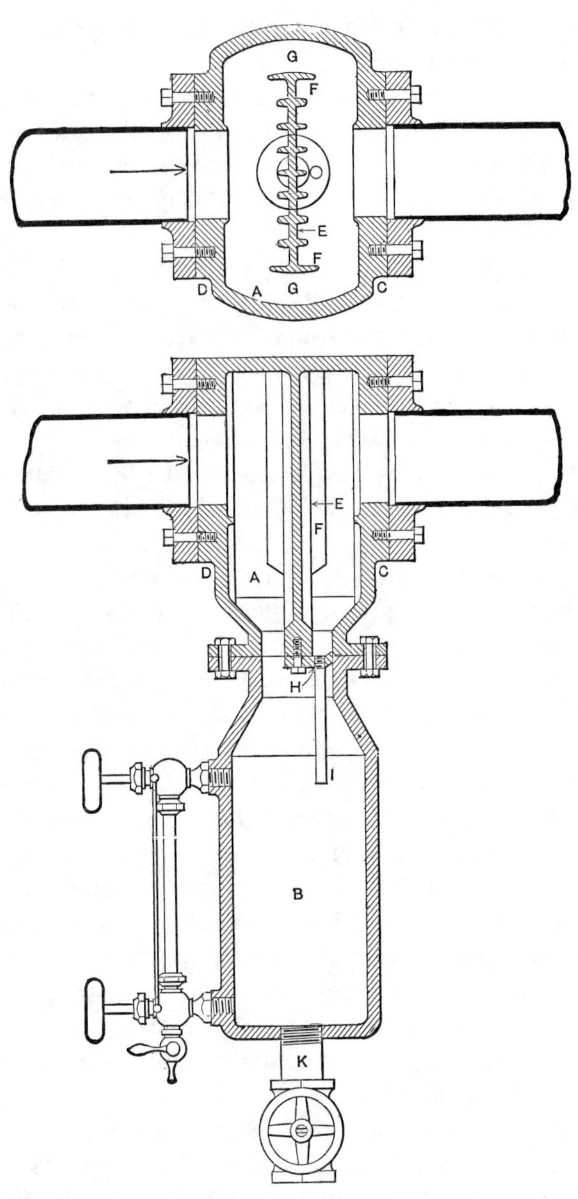

FIG. 107.—Cochrane Steam Separator.

FF. This plate E, which extends over the opening in C, is not carried entirely across the body of the separator. The passages GG allow the steam to pass through to the opening in D. At the bottom of E is bolted the small plate H, which prevents the steam passing directly into the reservoir B. The pipe I takes care of any moisture that may collect beyond the baffle-plate.

The wet steam entering from the left with a high velocity divides, part going to the front and part to the back, so as to pass the baffle-plate. The entrained water, however, is heavier, and having more momentum travels on, striking the baffle-plate. The ribs on the face prevent this water from being taken up by the steam, and it falls into B. From here it is conveyed through K to a trap. A gauge-glass is attached to B, so that, should any water collect, due to the failure of the trap to operate, it may be seen.

Special Valves.—Two special valves are often used in a steam plant when low pressure steam is desired, the reducing valve and the back pressure valve. The **McDaniel reducing valve,** shown in Fig. 108, consists of a **hollow piston** A, moving in a **casing** B. Steam enters at C, and passing through the triangular openings in the wall of the piston leaves at D. The pressure beneath the piston acts on the top and tends to lift the piston against the pressure exerted by the weight E through the lever F. The space above the piston is open to the atmosphere through the small opening, so that any steam which leaks past is discharged. If the steam pressure at D becomes higher than that for which the valve is set, the steam raises the valve and cuts off the steam-supply. The valve usually occupies such a position that the supply is just sufficient for the demand; should the conditions change the valve will alter its position.

When the system is such that the pressure is to be kept below a certain point, a form of safety-valve known as a back-pressure valve, Fig. 109, is used. The valve A of the **Jenkins' back pressure valve** is held down by the arm B, which is at-

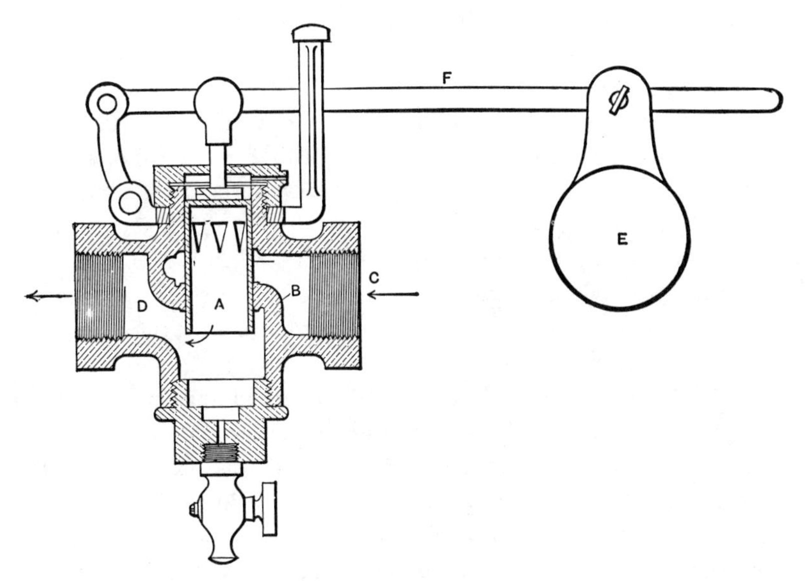

FIG. 108.—McDaniel Reducing Valve.

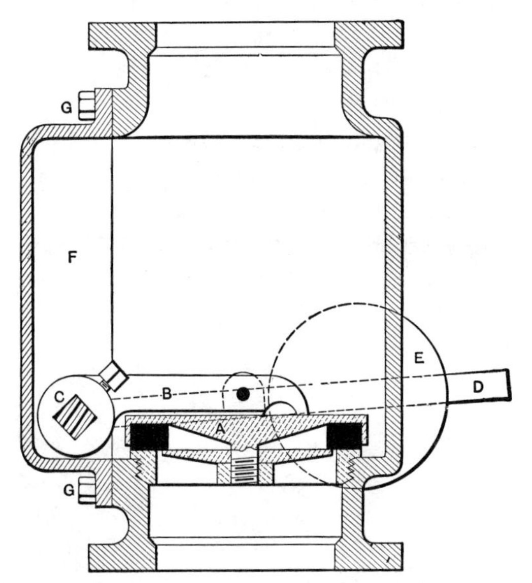

FIG. 109.—Jenkins' Back-pressure Valve.

tached to the shaft C. To C is attached the arm D, outside the casing, carrying the weight E. When the pressure below the disc is sufficient to raise the weight, the valve opens and discharges into the air. To repair the valve the disc and arm are removed, with the cap F, by taking out the tap-bolts GG. The valves are usually employed where pressures of a few pounds above the atmosphere are required, as in places where the exhaust-steam from engines is used for heating. It is important that when there is little need for the steam the back pressure on the engine be not increased, and hence the employment of such valves.

Water-hammer.—When steam is turned into a line containing water below the steam temperature, some of the steam may be condensed in such a way as to form a partial vacuum. This causes the water to be thrown in the direction of the reduced pressure, and there is set up an oscillation of this water, which increases to such an extent that much damage may be done. This, which is called water-hammer, is the cause of the pounding so often heard in steam-pipes. To guard against it, the steam lines are drained as described, and care is taken to prevent the accumulation of water when there is no steam on the line.

The method of making steam in a steam-boiler from fuels, and the construction of the boilers and the various auxiliaries which are used in the modern boiler-house have been described, and now the method of transforming the energy of the steam into useful work of some kind by the steam-engine will be examined.

CHAPTER IV.

THE SLIDE-VALVE STEAM-ENGINE.

STEAM is a vapor, and like all vapors presses on the walls of the confining vessel. This pressure is exerted equally in all directions, and, if the vessel varies in volume, the steam will do work. As the volume increases the pressure per unit area decreases, if the weight of steam is constant, but, as it still tends to further increase its volume, it is still capable of doing more work.

This action ordinarily takes place in the cylinder of a steam-engine, a simple form of what is shown in the plain slide-valve engine, Figs. 110 and 111.

The casting A, called the cylinder, is made of cast iron, and has extending through it a cylindrical surface forming the **bore,** the walls forming the **barrel.** The piston B, which moves in the bore, fits snugly against the walls so that no steam can leak between. The heads of the cylinder CC close the ends of the cylinder. The **steam-chest** (**valve-chest, valve-box**) D is formed on the side of the cylindrical casting in the form of a box, and the slide-valve E within it controls the admission of steam to the cylinder, steam entering the chest through the pipe F, Fig. 111. The steam pressure is transmitted from the piston through the **piston-rod** G, the **cross-head** H, and the **connecting-rod** I to the **crank** K. Fig. 111 shows how the pressure on the piston will turn the crank K and cause the **shaft** M to revolve, and with it the **fly-wheel** N and the **eccentric** O. The latter, which is equivalent to a crank, is connected to the **valve** E by the **eccentric-rod** P and the **valve-rod** Q, thus driving the valve back and forth. The valve E, containing the cavity R, is a block of cast iron with a finished face moving over a similar surface.

With the engine in the position shown, steam from the boiler, passing through the **steam-passage** S, connecting with

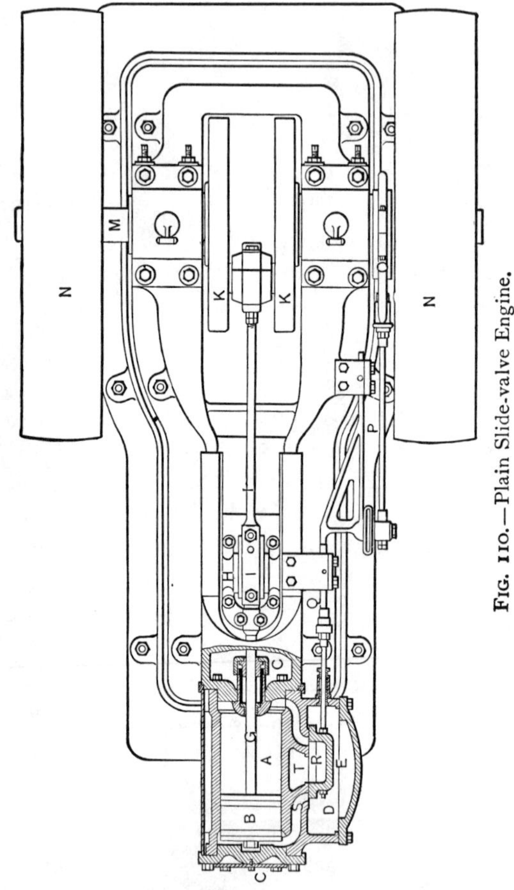

FIG. 110.—Plain Slide-valve Engine.

steam-chest, enters the cylinder to the left of the piston. The pressure of this steam causes the piston to move to the right. A cylinder 10″ in diameter with 100 pounds steam pressure has a force of 7854 pounds exerted on the piston, about $3\frac{1}{2}$ tons. As the shaft turns, the eccentric moves the valve to such a position that no more steam can enter the passage S, and the steam is said to be **cut-off.** The steam continues to move the piston, the pressure falls, and with it the temperature of the steam. In general this fall of pressure is nearly proportional to the increase of volume. When, for instance, the volume is

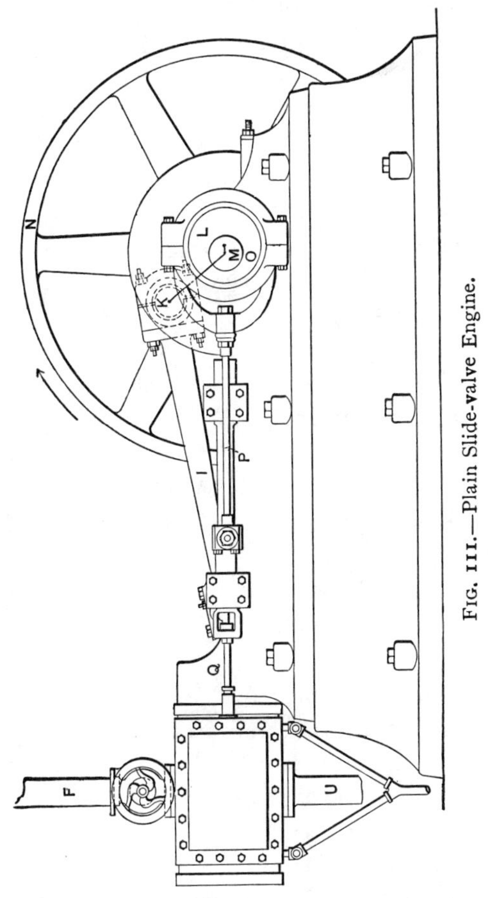

FIG. 111.—Plain Slide-valve Engine.

double that at cut-off, the pressure is reduced to about half its value at that point. If the pressure which resists the action of the piston is still lower than the steam pressure, the piston continues to move to the right. As this occurs, the eccentric moves the valve to the left, and, when the piston has almost reached the right-hand end of the stroke, this motion has placed the valve so that the expanding steam can escape through the cavity R to the **exhaust-passage** T, and thence to the air through the **exhaust-pipe** U. Although steam is admitted to the right-hand end of the cylinder before the end of the stroke, the inertia of the fly-wheel is sufficient to carry the piston to the end of the stroke against this pressure and allow it to begin the return. At the end of the stroke the crank and connecting-rod are in line, and the force from these rods does not tend to turn the crank, hence this position is spoken of as the **dead-center,** or **the center.** In case the engine stops on center it is necessary to move it off by some means before turning on steam. On the return stroke the action is the same as that just described, steam being admitted, cut off, and expanded on the right-hand side, while the exhaust of the steam on the left occurs through the exhaust passage to the air. This exhaust action is often discontinued before the end of the stroke, and then **compression** of the remaining steam takes place. This serves to fill the waste spaces with steam before the new steam is admitted. These actions are continued, the steam being automatically admitted and exhausted on one side or the other.

It may be asked how the engine is able to work when the pressure changes and reaches in many cases a lower value than the resistance to be overcome. It is for the purpose of permitting this variation in pressure that the fly-wheel is used. This wheel has a heavy rim, and during a portion of the revolution there is more energy received by the shaft from the piston, through the various parts, than it delivers to the apparatus which consumes the power. As a result of this the engine speeds up, using the excess of energy in increasing the kinetic energy of the fly-wheel. When the quantity received from the piston is below the demand, the slowing down of the engine draws this excess from the fly-wheel. Thus there is a constant change, during each revolution, in the speed of the fly-wheel, due to this storing and giving up of energy. As we make the fly-wheel heavier the change in speed for the same variation in steam pressure is smaller, and hence the engine runs more uniformly. It is necessary, therefore, to have heavy fly-wheels on engines which are intended to drive machinery requiring very steady motion.

The engine shown in Fig. 111 is said to run over as the motion from the **head end** of the cylinder to the **crank end** occurs while the crank is above the engine center-line. The **crank end** of the cylinder is that nearest the crank. The amount of motion of the piston is called the **stroke** of the engine, and this is twice the distance from the center of the crank-pin to the center of the shaft, which distance is called the **crank radius,** or the **throw** of the crank. In giving the dimensions of an engine it is customary to mention the diameter of the cylinder, then the stroke in inches, and the number of revolutions of the engine per minute. Thus a 6×10—250 R.P.M. would mean an engine with a cylinder 6 inches in diameter, with a stroke of 10 inches, which turns two hundred and fifty times per minute.

CHAPTER V.

ENGINE DETAILS.

Cylinders.—The cylinder of an engine is made in many forms, as the various parts are differently arranged. Fig. 112 shows the construction of a cylinder used on a 6×8 Weston engine. The cylinder is made of cast iron on account of its complicated form. It is closed at the ends by the heads AA, forming a closed vessel into which steam enters from the steam chest by the passages BB. The steam-chest is cast with the barrel of the cylinder, and is covered by the **steam-chest cover** C. The barrel of the cylinder is carefully bored out to a uniform diameter. Near each end the diameter is increased, forming the **counter-bore.** The piston D contains rings E, which project beyond its body and press against the cylinder. These rings would gradually wear the cylinder and form a projection or shoulder at each end, which would cause damage should the extremities of the travel be altered by wear or by change in the length of the connecting-rod. To guard against this, the counter-bore is made so that the piston-ring travels a short distance beyond the main bore, but not far enough to allow the ring to slip out. This arrangement prevents a shoulder forming, and should be employed in all cases where one body slides over another. The heads of the cylinder are cast-iron plates, often made with projections extending into the counter-bore of the cylinder. They are cast to conform to the outline of the piston and nut, and are held to the cylinder by stud-bolts and nuts, the joint being made in any one of several ways. For instance, the cylinder-head has its **flange** F turned to fit the flange of the cylinder, and then, after oil and fine emery have been placed between, the head is turned, grinding the two surfaces together. After this the head is bolted down and a steam-tight joint results. Again, a sheet of

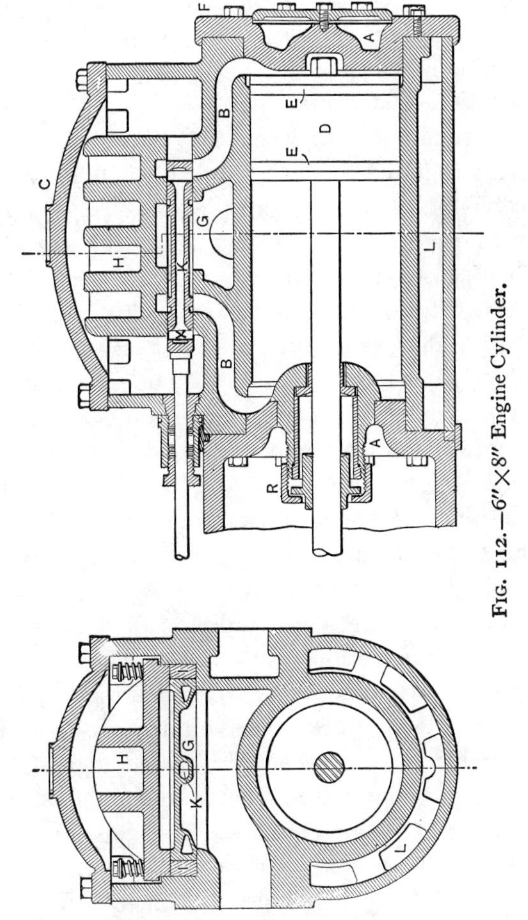

FIG. 112.—6″×8″ Engine Cylinder.

thick paper or rubber gasket may be cut to fit between the two, which, when the head is bolted fast renders the joint tight. A piece of copper wire or asbestos rope may be placed between these surfaces, and this on being flattened by the pressure makes a satisfactory joint.

The steam-chest cover C, Fig. 112, has a steam-tight joint

made by a gasket and studs. Within the chest is the valve G, moving under the pressure-plate H. If the steam were to act on the back of the valve the pressure on the seat would be excessive and give rise to undue friction and wear. This pressure, even in small engines, may be as much as a ton. To prevent loss by friction the pressure-plate H is used which rests on **side-blocks,** or **distance-pieces,** I of such a size that the valve has a running fit between the plate and seat. In this way the force required to move the valve is greatly reduced.

The valve G is made flat, and contains a passage K so made that as steam from the steam-chest passes the right-hand edge of the valve it also enters the cavity X on the left hand, and thence through the passage K to the cavity on the right. At this point it discharges into the steam-passage, with the steam passing the right-hand edge. It will be seen that for a short distance the amount of opening for steam is twice that due to the valve movement, this valve being called a **double-ported valve.**

The distance between the piston and cylinder-head is termed the **clearance distance.** This is provided to allow for any inequalities in the two surfaces, or any slight errors in the lengths of the parts, and also to permit the change of the piston position as the various parts wear. The volume of this space, together with that of the steam-passage as far as the valve face, is called the **clearance volume,** or simply the **clearance.** The clearance is measured in percentage of the volume swept through by the piston (the **piston displacement**), and varies from two to fifteen per cent. It is possible to compress some of the exhaust-steam in this space at the end of the stroke, thus making a **cushion** which assists in bringing the parts to rest, and it is then not necessary to fill the clearance space with boiler steam at each stroke. This volume also permits of a small amount of water being in the cylinder, which might force off the head were it not for the clearance.

In Fig. 112 the cylinder is cast with a hollow space L extending around it. This space is called a **jacket.** In the

Weston engine it contains air, but engines are built in which this is filled with steam. The **steam-jacket,** which also may be formed in the hollow heads as well as in the **walls** of the cylinder, is used to keep the temperature of the cylinder more nearly uniform, and thus reduce the condensation of the steam as it enters the cylinder. As stated before, when the steam expands after cut-off the temperature falls, reaching, at exhaust, a low value. This of course cools the cylinder walls, and when steam is brought into the cylinder some of it is condensed as it strikes the cold walls. This **initial condensation,** which may amount to forty or fifty per cent of the steam supplied, is a source of great waste, and to reduce it the jacket is used.

It is difficult to cast a jacket on a large cylinder as the unequal cooling of the parts is apt to crack some portion of the

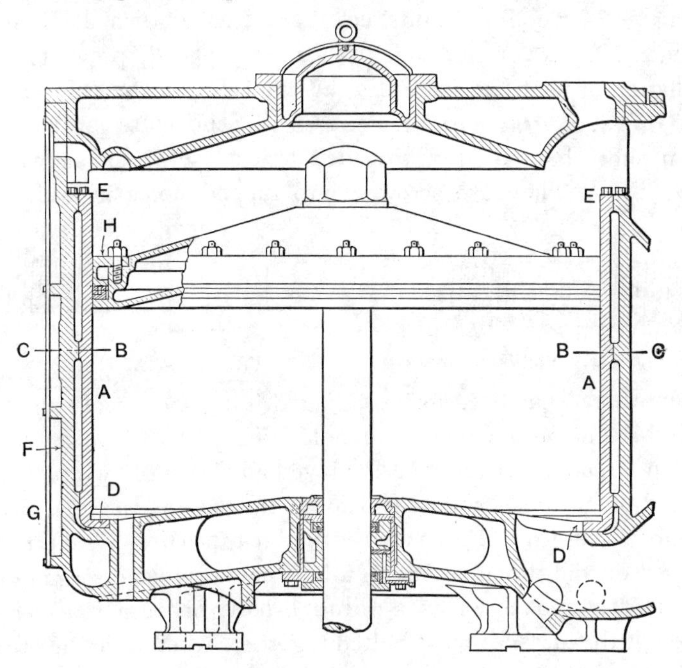

FIG. 113.—Marine Cylinder.

casting. To overcome this difficulty, and also to have a cylinder of better mechanical construction, liners are used. These are cylindrical castings A, shown in Fig. 113, which have turned projections B on their outer surface fitting on rings C which project from the inner surface of the main cylinder and which have been bored to fit the projections on the liner. At the bottom a flange D turns inwards, and is screwed by tap-bolts to the barrel, while at the top, to allow for expansion, the joint E is made by a piece of sheet copper held beneath two rings. The main cylinder can be made of soft iron, while the liner is made of a tougher grade, and, in place of a complicated casting, we have a plain barrel. This liner may be renewed when necessary.

The figure also shows a covering F, placed around the cylinder, which consists of some non-conducting substance, such as hair-felt, asbestos, magnesia, or cork, outside of which is a covering G of wood or **planished sheet iron,** known as **Russia iron.** This covering, which is called **lagging,** prevents the radiation of heat and increases the efficiency of the engine.

Bolts, Nuts, and Wrenches.—**Stud-bolts** are made from short bars of iron threaded at each end, as shown in Fig. 114. They are screwed into tapped holes in which a

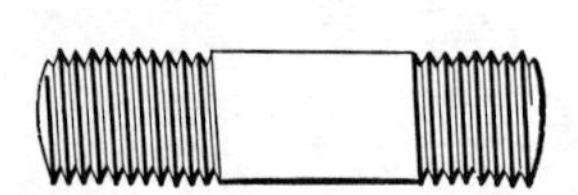

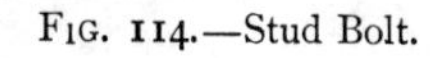

FIG. 114.—Stud Bolt.

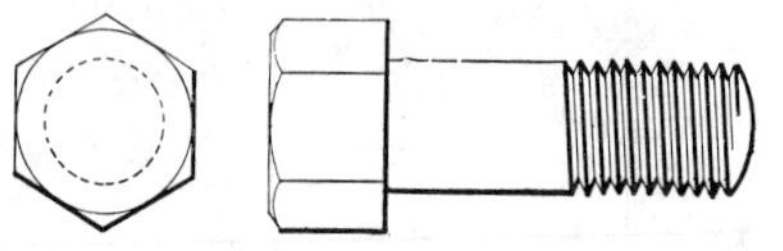

FIG. 115.—Tap Bolt.

screw-thread has been formed, after drilling, by a tool called a tap. A nut on the outer end holds the piece to be secured. When a stud-bolt cannot be employed a bolt is used with a head forged on one end to take the place of the nut and thread. Such a bolt, as shown in Fig. 115, is called a **tap-bolt.** When a nut is used on the threaded end of a bolt having a head forged on, the bolt is spoken of as a **through-bolt,** because it extends through the surfaces to be bolted together, whereas in the other forms the bolt screws into one of the surfaces. The through-bolt, Fig. 116, is the one ordinarily spoken of as a **bolt.**

The bolt-head is sometimes made as shown in Fig. 117,

when it is said to be T-headed. This is used when the head of the bolt is to be passed into a slot as shown.

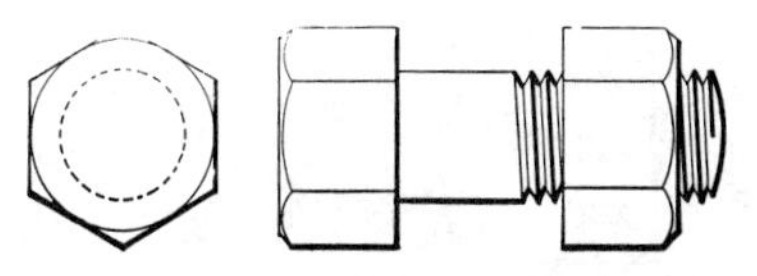

FIG. 116.—Through Bolt.

The nuts which screw on the bolts are of different forms, that shown in Fig. 116 being the ordinary hexagonal nut. Fig. 118 shows the form of nut used when it is desired to cover

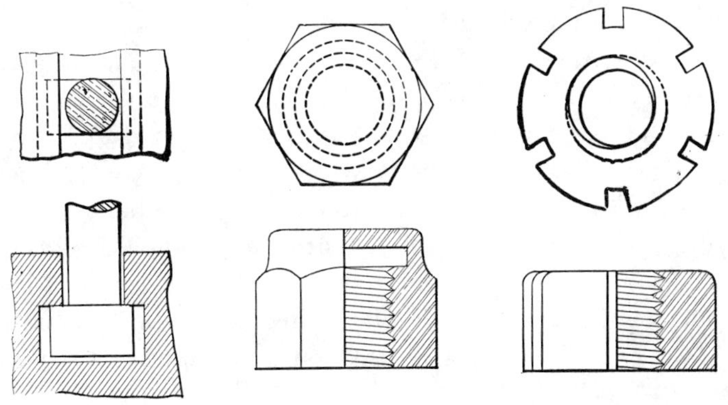

FIG. 117.—T-headed Bolt. FIG. 118.—Cap Nut. FIG. 119.—Fluted Nut.

the end of the thread to prevent leakage, as on the end of a boiler stay bolt, or for finish, as on cylinder heads of some engines. Circular nuts are often used where it is not convenient to use a hexagonal one. These nuts are either fluted, as shown in Fig. 119, or are pierced around the periphery by small circular holes perpendicular to the axes.

To tighten the nuts various wrenches are used. The **monkey wrench,** Fig. 120, consists of a fixed jaw A at right angles with the bar B on which it is formed. The wooden handle C is fastened to the end of this bar. The screw C is attached to projection D from the bar B, so that it may screw into the nut F on the movable jaw E, this moving it on B and making the wrench adjustable. The nut has its edge serrated, or **knurled**, to give a better finger grip.

This adjustable principle is employed in various wrenches for nuts and pipes. In the latter case the adjustable jaw has a

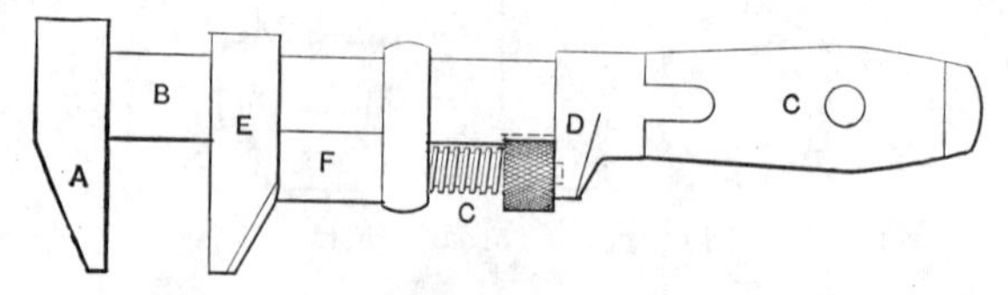

FIG. 120.—Monkey Wrench.

slight movement about an axis on the moving nut E, and the edges are serrated so that it may grip the pipe. Such a wrench is known as a **pipe wrench.**

The objections to the monkey wrench are that on heavy work the jaws are apt to spring and thus damage the nut, and the size of the jaws prevents their use when the nuts are located in corners. The **solid wrench,** Fig. 121, overcomes these objections. In these wrenches, the jaws A and handle B are made in one piece, and usually are **drop forgings,** although the larger sizes are made by machine and hand tools. The jaws are inclined at an angle of fifteen degrees to the line of the handle, when the wrench is used on a hexagon nut, so that

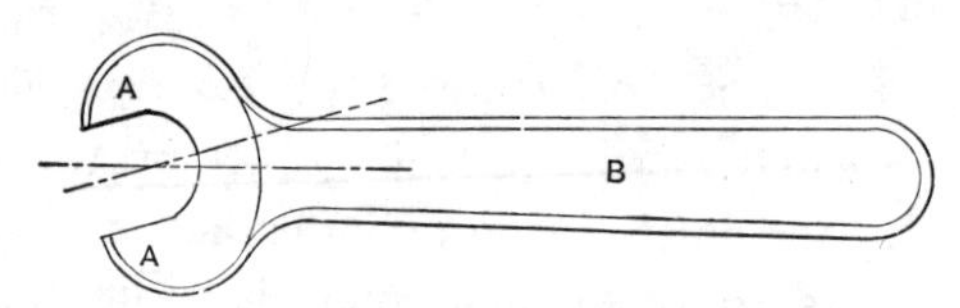

FIG. 121.—Solid Wrench.

the wrench may be used when the nut is so situated that it is only possible to move it, by the wrench, through thirty degrees. In such a case, by turning the wrench over, the nut can be moved through another thirty degrees and the nut faces are again in their first position. With the axis of the jaws parallel to the axis of the wrench, the reversal of the wrench would accomplish nothing. To reduce the number of separate wrenches necessary on machines, jaws are often placed at each end of the handle, making an **S wrench.**

The spanner wrench, Fig. 122, is used on circular nuts,

the projection being a lip or pin which fits the fluting or hole. To reverse the motion of the nut the spanner wrench is reversed. A form of wrench used when there is not sufficient room to use one of the forms just mentioned is the **socket wrench,** Fig. 123. In this wrench a cavity A is made on the end of a bar to fit the nut. The wrench is turned either by using an ordinary wrench on B or by a pin through it, or by a T-handle forged on

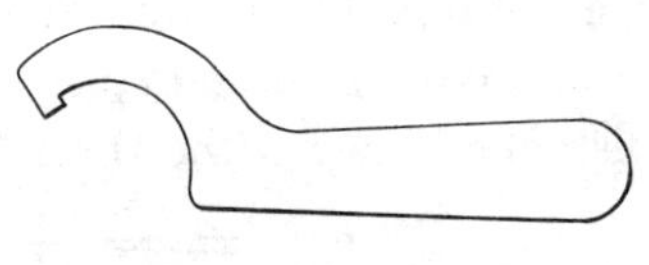

FIG. 122.—Spanner.

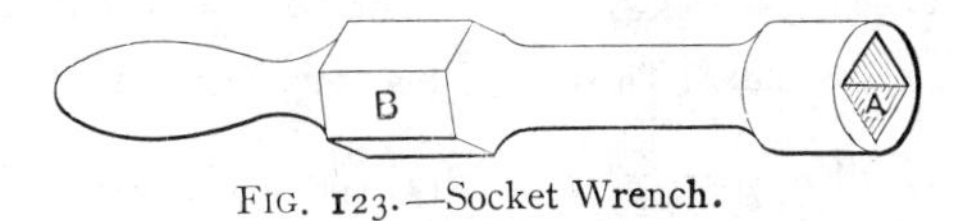

FIG. 123.—Socket Wrench.

The threads on the bolts in Figs. 114 to 116 are made with slanting sides, while those shown on the monkey wrench, Fig. 120, are square. The first is the **Sellers',** or **United States standard** thread, in which the threads are made with sides making an angle of sixty degrees and with one-eighth of their height cut off at the top and filled in at the bottom. The square thread has its sides square with the axis, and the thread is of the same width as the space. The first of these, Fig. 124, is used on bolts fastening bodies together because of its strength and ease of manufacture, while the latter, Fig. 125, is used where the screw is used to transmit power, or is to be used in a place where it may be roughly handled. Modifications of these threads are sometimes used.

The lines of the thread of the bolt form what are known as helices. A **helix** is a curve on the surface of a cylinder, which as it extends around the cylinder uniformly rises, so that when it reaches the element of the cylinder from which it was started it is above the point of starting. The distance between these successive crossings is called the **pitch** of the helix. When the thread is so arranged that there is only one helix on the bolt, the bolt is said to have a **single thread;** but often in square threads there are two helices, so that the distance between threads is one-half the pitch of the screw. Such a bolt is said to have a **double thread**. This form of thread is used to make the nut travel faster and at the same time to retain the strength of the bolt, as the depth of the thread is proportional to the distance between the threads.

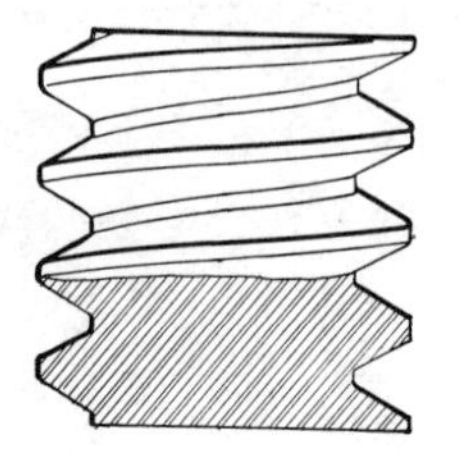

FIG. 124.—Standard Thread.

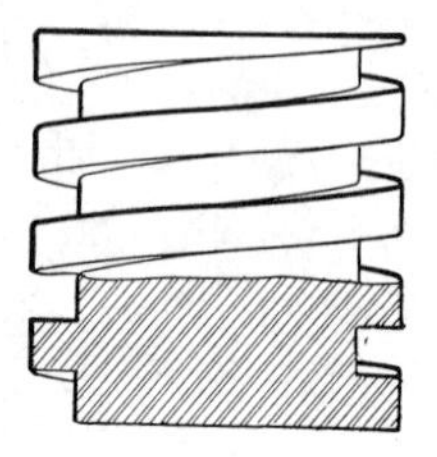

FIG. 125.—Square Thread.

Piston.—The piston is a shallow cylindrical casting of iron, made deep enough for strength, and sometimes to prevent the steam from passing around it. It is generally cast hollow to reduce the weight, and, in large sizes, webs are cast in this space to stiffen the flat surfaces. The piston shown in Fig. 126 is similar in construction to many in use. The piston rod A, which is made of steel, is turned to a taper on the piston end, so that it may be made tight without much driving, which would be required with a parallel rod properly made. The shoulder B determines the position of the piston on the rod, and prevents excessive pressure being produced by drawing up the nut C. The nut on the end is **locked** in place by a pin D, which extends through the nut and rod. At times a **lock screw** is used which is placed in the end of the nut and rod, so that one-half of the screw is in each.

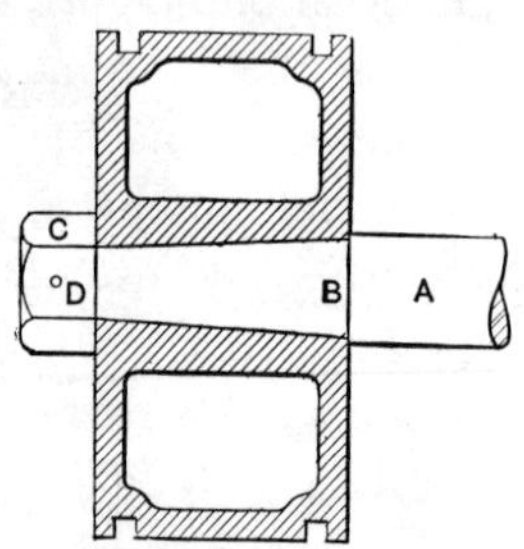

FIG. 126.—Piston.

In some cases the piston is shrunk on the rod, as shown in Fig. 127, the piston being expanded by heating and the end of the cold rod slipped in. The rod is turned to a diameter slightly larger than the hole through the piston, and is held tightly when the piston contracts on cooling. The rod is then riveted over at the end. The screw plug A is a **core-plug** to close the opening into the hollow head after the core sand has been removed.

To make the surface between the cylinder and piston steam tight, **piston packing rings** are placed in grooves in the casting. These are often made of cast iron, B, Fig. 127, turned

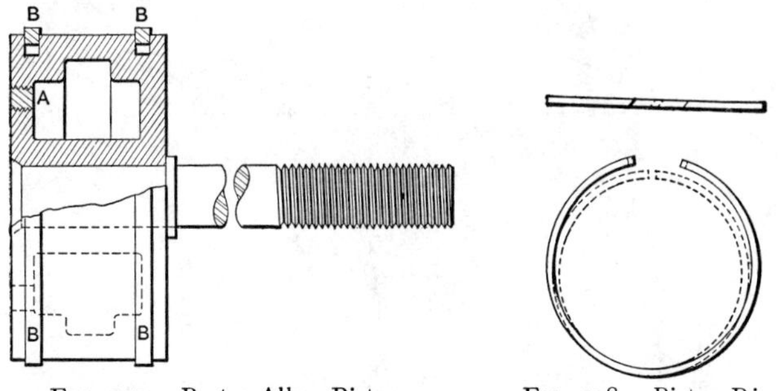

FIG. 127.—Porter-Allen Piston. FIG. 128.—Piston Ring.

to a diameter larger than the cylinder bore. After a portion of the ring has been removed the ring bears lightly on the cylinder wall on being sprung in place. Fig. 128 shows the construction of a simple ring with such a section cut out that when the ring is sprung into the cylinder the ends almost meet. These rings are generally made of uniform cross section, having their outer surface and sides finished. The inner surface is sometimes left rough. A ring of variable section is employed at times, although not as extensively as that of uniform section.

Figs. 113 and 129 show the method of constructing and packing a marine piston such as is used in the U. S. Navy. The piston shown in Fig. 129 is made of cast steel, with a conical body A and heavy rim B to make it rigid. The piston rod C is forced into the piston boss D and secured by a large nut E. The lower end F of the rod is finished to fit the cross head. Two piston rings G are placed together, and are so arranged that their joints are not at the same point. In such an arrangement the rings are said to **break joints.** The object of this is to prevent the steam from blowing through the two openings, which would occur were they in line.

The hollow piston shown in Fig. 113 is made of composi-

tion (brass), and is packed by a three-ring packing, as shown.

When a single wide ring is used blowing through is pre-

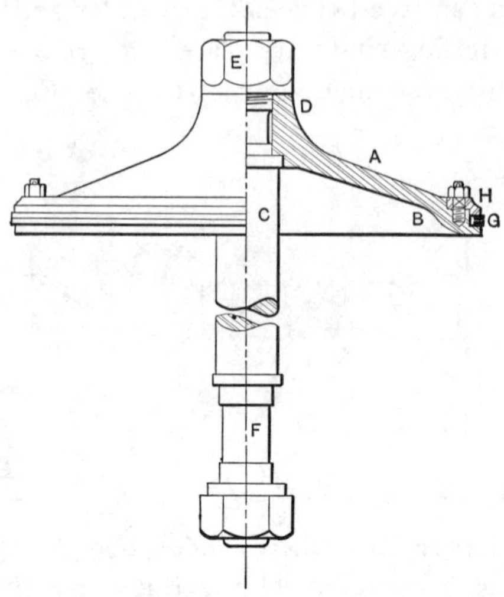

FIG. 129.—Marine Piston.

vented by the use of a tongue, as shown in Fig. 130. This allows the ring to expand, but at the same time prevents leakage. The piston packing-rings, Figs. 113 and 129, are held in place by the **junk-ring,** or **follower-ring,** H. This ring, which is held by a number of studs, makes it a simple matter

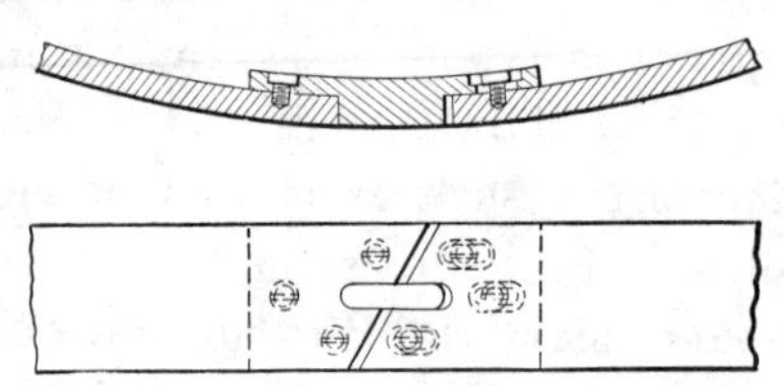

FIG. 130.—Piston-ring with Tongue.

to inspect or replace packing-rings. The term junk-ring recalls the former practice of packing with hemp rope soaked in tallow, which was called junk. Such a method was suitable only for the low pressure used in the early days of steam-engineering.

The piston-rings are sometimes made in sections, an example of which is shown in Fig. 131. The ring is made of small sections AA, each of which has machined ends BB, which nearly meet the ends of the adjacent sections. The joint is made tight by the keeper C, which fits over the ends of the sections. The spring D behind the keeper, and H behind the section at the middle, hold the rings against the cylinder walls. These sections are carried in the **bull-ring** E, which is centered in the cylinder by the set-screws F in the body of the piston. The

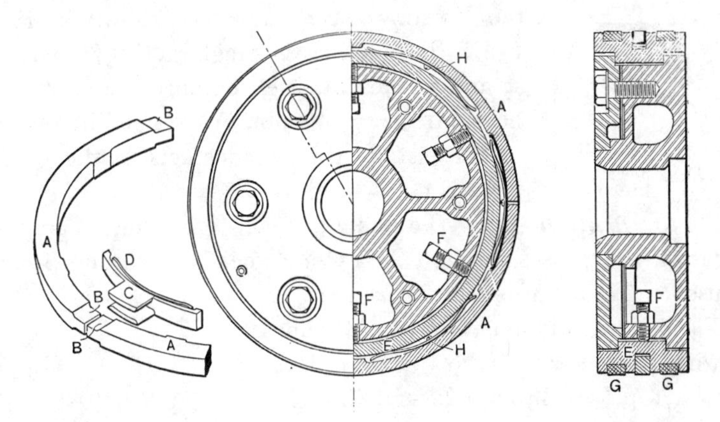

FIG. 131.—Sectional Packing-rings.

rings GG are solid, and are made of a hard composition fitted into the bull-ring, and are intended to take the weight of the piston. The section-rings have, therefore, only to make the joint steam-tight.

The piston-rings of all pistons are fitted so that they may easily move in their grooves, but yet be tight enough to prevent steam-leakage. This allows the ring to remain tight against the cylinder wall, although the piston may move away from a central position. In the case of large pistons in horizontal engines, the groove may be turned eccentric to the outer circumference of the piston, so that while at the bottom the ring is in contact with the piston-casting and carries the weight, the top is free to move. This is also accomplished by placing a block on that portion of the ring which is at the lowest point of the piston. The weight of the piston is apt to wear the cylinder at the lowest point. To prevent this the piston-rod is sometimes continued through the rear head, thus carrying the weight of the piston on the rods and bringing the wear on the supports of the rod instead of the cylinder. Such an extension of the piston rod is called a **tail rod.**

Trunk pistons.—In certain engines the piston, piston rod, and cross head are combined in one piece, forming a **trunk piston.** Such a piston from the Westinghouse engine is shown in Fig. 132. This piston is **single acting;** that is, steam is taken on one side only. The **cross-head pin,** or **wrist pin,** A, is held in bosses on the piston; the cylinder acts as the guide for this cross head.

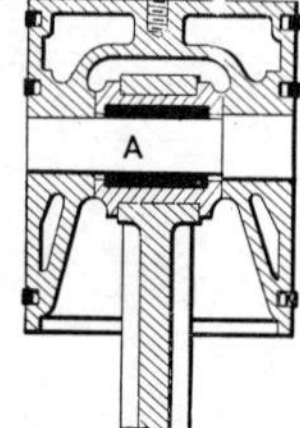

FIG. 132.—Trunk Piston.

Stuffing boxes.—To make a steam-tight joint where the piston-rod passes through the cylinder heads a stuffing box is used. This consists of a casting A, Fig. 133, containing a cavity B somewhat larger than the piston rod C, which passes through it, and in which material of an elastic nature is forced against the rod, by a cap or **gland,** D. The cavity is closed at one end by a brass **bushing** E, and at the other by the brass gland D. In many engines the brass bushings are omitted, the iron of the cylinder and gland being bored out properly to form the stuffing box. The brass bushing has the advantage that it can be more easily made a good fit, and is more easily renewed.

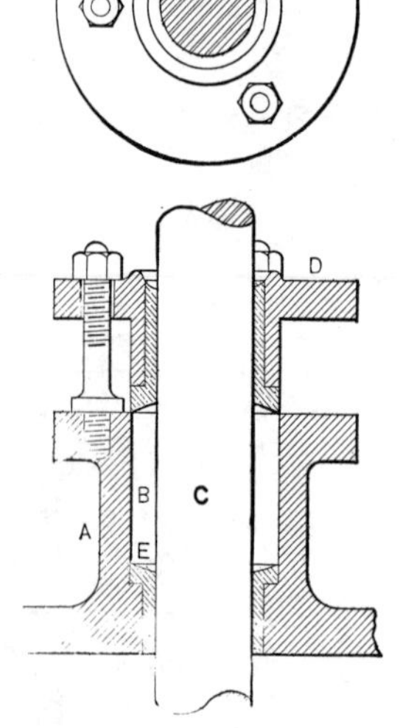

FIG. 133.—Stuffing Box.

The material called **piston-rod packing,** used to make this joint steam-tight, may be made of woven strands of hemp or cotton, with or without a central core of india-rubber, or asbestos rope may be used.

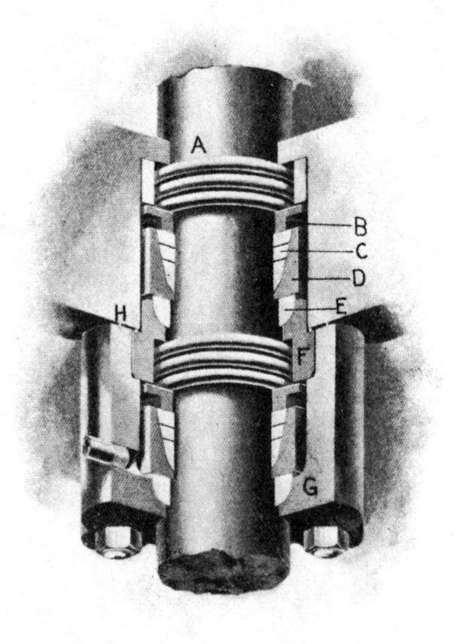

FIG. 134.—United States Metallic Packing.

(*To face page* 125.)

The stuffing-box shown in Fig. 133 is cast with the cylinder-head, that shown in Fig. 113 has a separate box, which is fitted with metallic packing. The stuffing-box shown in Fig. 112 is made separate from the cast-iron cylinder-head, and is intended to be packed with a soft packing.

The method of tightening the gland is quite different in Fig. 112 from that shown in Fig. 133. In the former case the cap R forces the gland down, while in the latter three nuts are used on stud-bolts. For large glands the nuts are so connected by gears that they turn simultaneously, causing the gland to enter squarely.

Metallic Packing.—Many engines are now equipped with metallic packings in place of the fibrous forms. These packings will automatically adjust themselves as wear takes place, and are so flexible that they will allow the piston-rod to move out of its central position.

Fig. 134 shows the construction of the **United States metallic packing.** The bottom of the stuffing-box contains a spring A, which, acting on a cast-iron follower-ring B, presses the three babbitt-metal rings C against the conical cup D, and so tightens them against the rod. At the end of the cup D is a spherical ring E, which fits into the ring F, containing a spherical cavity, forming a **ball and socket joint.** This ring F forms the basis of a second set of packing, the upper ring of which fits the cap G of the stuffing-box. By this arrangement the axis of the piston can be displaced without any leakage, and by the ball and socket joint, the alignment of the axis is not fixed. In large engines having considerable cross-motion of the piston-rods no leakage occurs. The joints between the cap and the cylinder-casting are made by means of copper wire HH.

Babbitt metal is a soft alloy of copper, tin, and antimony. This and other alloys, generally known as **white metal,** are used where sliding occurs, the wear taking place on the soft alloy rather than on the rods. They have the advantage that the wearing surface can be readily replaced or put in good working shape after being injured.

Valves.—The various forms of valves controlling the action of the steam in the cylinder will be considered in detail later, but there are several other valves that are necessary for the proper action of the engine.

Throttle or Stop Valves.—The steam pipe leading to an engine is supplied with some form of valve similar to those described in Chapter II. This valve is used to shut off the supply of steam, and is therefore termed the stop valve. Large stop valves for engines, or for other purposes, whether gate or globe valves, are supplied, usually, with a small connection leading from one side of the valve to the other, known as a **by-pass.** This connection contains a valve, so that, while the main valve is closed, steam can be admitted to the cylinder to warm up, or to raise the pressure on the low pressure side of the main valve that it may be the easier to open. On marine engines there is usually another valve supplied, known as the **throttle valve.** This consists of a diaphragm mounted on an axis passing through the center of the pipe. When this diaphragm is turned so that it lies in the axis of the pipe there is little resistance to the passage of the steam; but, as it is turned, it gradually closes the pipe and checks the supply of steam. When completely closed it fills the pipe, allowing very little steam to pass. The **trunnions** on which the valve turns pass through stuffing boxes, and the handle for controlling the valve is attached to one end. Such a valve is termed a **butterfly valve.** The term throttle valve is often applied to stop valves.

Pet Cocks.—To drain from the cylinder the water which may collect while the engine is stopped, small valves are placed at the lowest points of each end of the cylinder. When these are open, the water can drain off, and, on starting, the steam blows out the water that remains. These valves are usually called **drips, pet cocks,** or **drain cocks.** A **relief valve,** Fig. 135, is a valve, similar to a spring safety-valve, which is placed on each cylinder head, so that, should there be a large amount of water in the cylinder, it would be discharged through the relief before the pressure could become great enough to tear off the head. The spring A is so adjusted that it will not raise until the pressure greatly exceeds that of the boiler steam. The valve B is guided by the wings C, and held in place by the spring A, which may be adjusted to the required pressure by the set screw D, which is secured from turning by the jam-nut E.

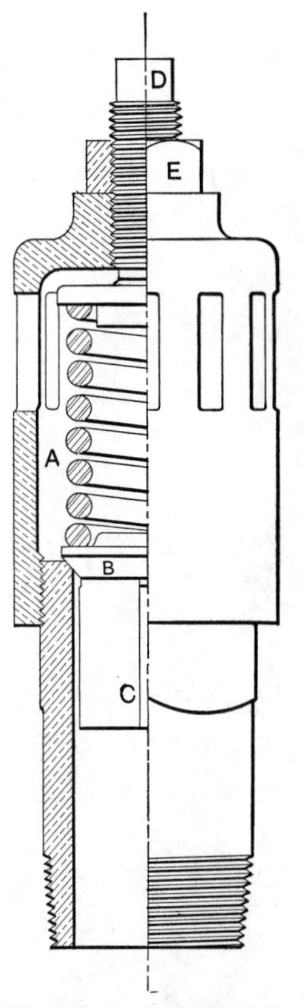

FIG. 135.—Cylinder Relief Valve.

Before passing to the consideration of the cross head and connecting rod it would be well to examine the forces acting on these parts. In the upper part of Fig. 136 the force from the piston produces compression in the rod A and drives the cross head to the right, which with the resistance from the crank, produces compression in the connecting rod B. This resistance from the connecting rod, since it is not in the same line with A, must produce a downward pressure on the cross head. This pressure on the guide acts throughout the whole stroke, reaching a maximum value near the middle. On the return stroke, as shown in the lower diagram, the rod A is in tension, as is the rod B, the crank C being pulled by it. In this case, also, the resistances from the connecting rod cause a pressure on the lower guide. When, however, compression takes place, the fly wheel may drive the piston instead of the piston driving the fly wheel, and the action in Fig. 136 is reversed, and the cross head is forced against the upper guide. This usually occurs near the end of the stroke, where the pressure on the guides is small. For this reason the upper guide is necessary, but its surface need not be as large as that of the lower one. If the direction of the engine is reversed, the guide pressure would act on the upper surface.

Cross Heads.—The piston rod is attached to the cross head, from which the connecting rod is driven. In each of the

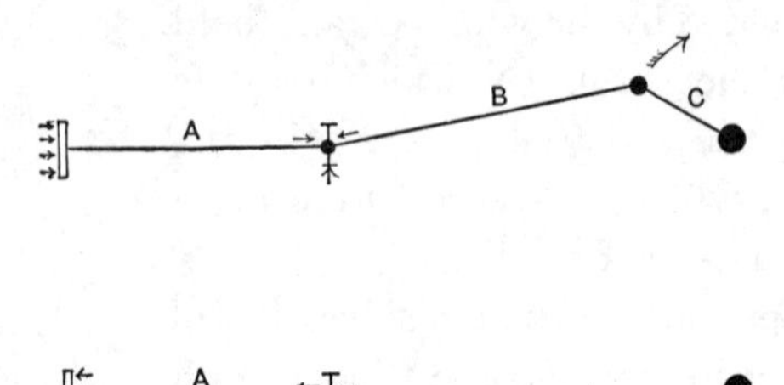

FIG. 136.—Forces on Piston Rod, etc.

following examples of cross heads, there is a cross-head pin **A**, a sliding surface, and a part C for the attachment of the piston rod.

Fig. 137 shows the cross head used on the Porter-Allen

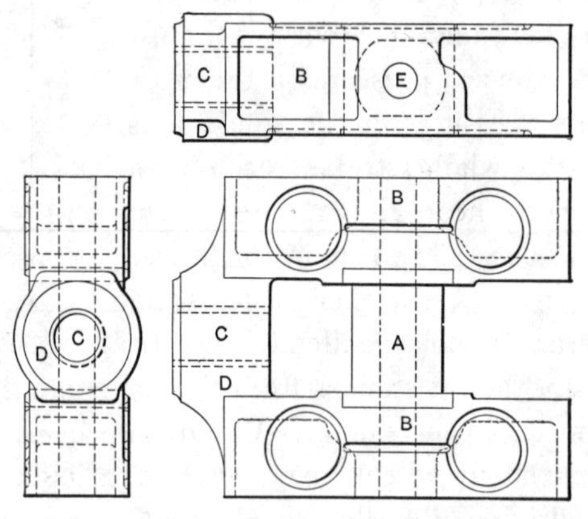

FIG. 137.—Porter-Allen Cross Head.

engine. The slide blocks BB are united by the yoke **D**. This yoke contains a threaded hole into which the piston rod is screwed. To prevent the rod from backing out, a **lock nut** or **jam nut** is screwed up against the yoke.

When the bolt is subject to vibration, or is so placed that the load is taken off at times, it is liable to "**back off,**" because when the load is removed the looseness of the fit may allow the bolt to turn. To obviate this another nut is screwed on, which always keeps the pressure on the threads on one side and so prevents turning.

Fig. 138 shows the end of the piston rod A, part of the cross head B, and the lock nut C in place. Without the lock nut the load tends to turn the rod in the yoke, being prevented only by the screw friction, and a reversal of the force or vibration of the machinery increases this tendency. When, however, a lock nut is screwed down upon this yoke it pulls the rod upward so that the threads of the rod are in contact on the upper surfaces with the yoke, and on the lower surfaces with the lock nut. In this way the rod is so jammed between them that the greatly increased friction prevents it from turning. In using a lock nut on an ordinary bolt two nuts are used, and, after the lower nut is tightened, screwing down the upper nut brings the two tightly together and puts the load on the upper nut, and for this reason the upper nut should be the larger. It is easily seen that these nuts will not turn, even if there is no load on the bolt.

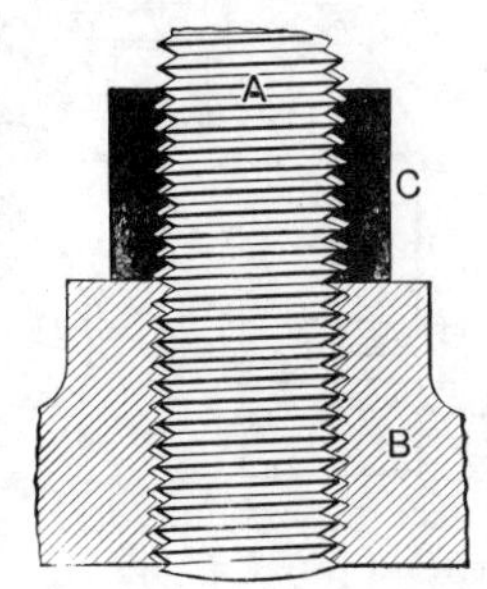

FIG. 138.—Lock Nut.

The cross-head pin A, Fig. 137, is made of steel, the body of the cross head being made of cast iron. The cross-head pin is held in place by a small pin E, which carries the vertical component of the connecting-rod thrust. The horizontal pressure from the connecting rod is carried on the square flanges at each end of the pin. To keep the piston rod in line and to carry the thrust produced by the **angularity** of the connecting rod, it is necessary to support the cross-head by **guides.** The blocks, together with the **guide surfaces** over which they slide are planed, and in some instances scraped, to reduce the friction. In the form of cross head shown, the blocks are provided with grooves in their surfaces to aid lubrication. In many other cross heads the block is faced with some form of Babbitt metal, or a brass piece, called a **shoe** or **slipper**, is fastened to the casting to form the sliding surface.

Fig. 139 is an example of a cross head used on a Corliss

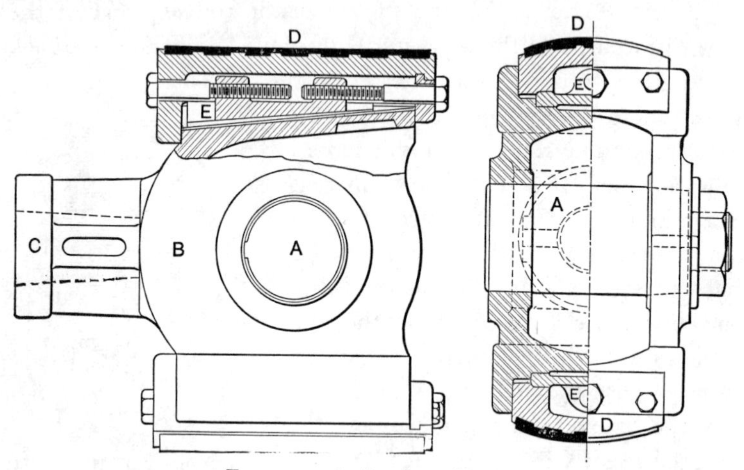

FIG. 139.—Corliss Cross Head.

engine. In this cross head the body B is of cast iron and the pin A of steel. Shoes D, faced with Babbitt, are attached to the body of the cross head in such a manner that, as wear takes place, they can be drawn up by means of wedges E. The piston rod is turned tapered and has a key way cut through it, entering a hole in the cross head and being held tight by the key. The round surface of the shoe is turned in a lathe, and, by means of a boring bar, the guide surfaces, which are usually cast in the frame of a Corliss engine, are bored.

A locomotive cross head is shown in Fig. 140. In this cross head the cast-iron body B contains the steel pin A and the tapered hole C for the piston rod, which is held by a key. There is only a single guide bar in this example, the shoes D and the side plates E enveloping it completely. The cross-head pin A is tapered where it passes through the cross-head casting, in order that it may be made tight without driving. The shoes D in this form of cross head are called **gibs.** F is

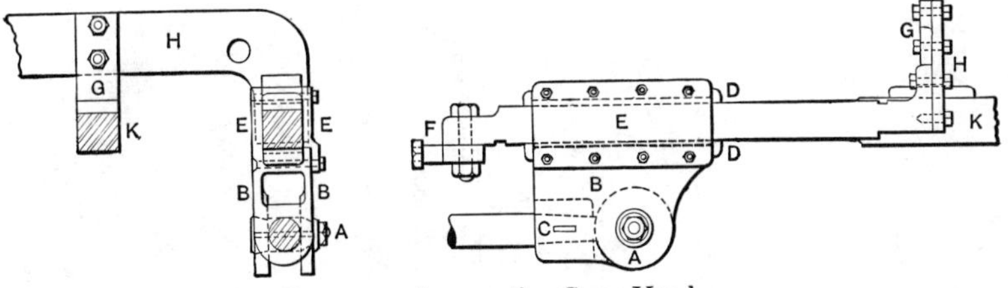

FIG. 140.—Locomotive Cross Head.

the attachment to the cylinder casting, and G is the attachment to the **guide bearer** H on the engine frame K.

The cross head of a marine engine, Fig. 141, is made with a portion C, solid with the piston rod. In this the cross head pin A, which is part of the connecting rod, works. The shoe B, which is bolted to this central portion, has its sliding surface formed of small blocks of Babbit metal inserted in cavities cast in the slide D. Plates E are bolted to the guide surface inclosing the shoe, forming the upper guide. This guide is brought into use when the engine is reversed, as well as at the end of the stroke. The arrangement of the Babbitt metal in squares makes lubrication simple. To keep the surfaces cool when the load is excessive, sea water is circulated through the guide casting at F.

In many marine engines the cross head carries two pins, over which the forked end of the connecting rod fits. Such pins are called **gudgeons.**

These cross heads show different arrangements of **guide bars.** In the first case, Fig. 137, four bars are used, two of which are formed by the **frame** of the engine, or **housing;** for the second, Fig. 139, two guide bars are used, both of which are formed by the engine frame; while for the third and fourth, Figs. 140 and 141, only one bar is used. The third case shows the cross head enveloping the guide bar, while in the fourth case the guide bar envelops the cross head.

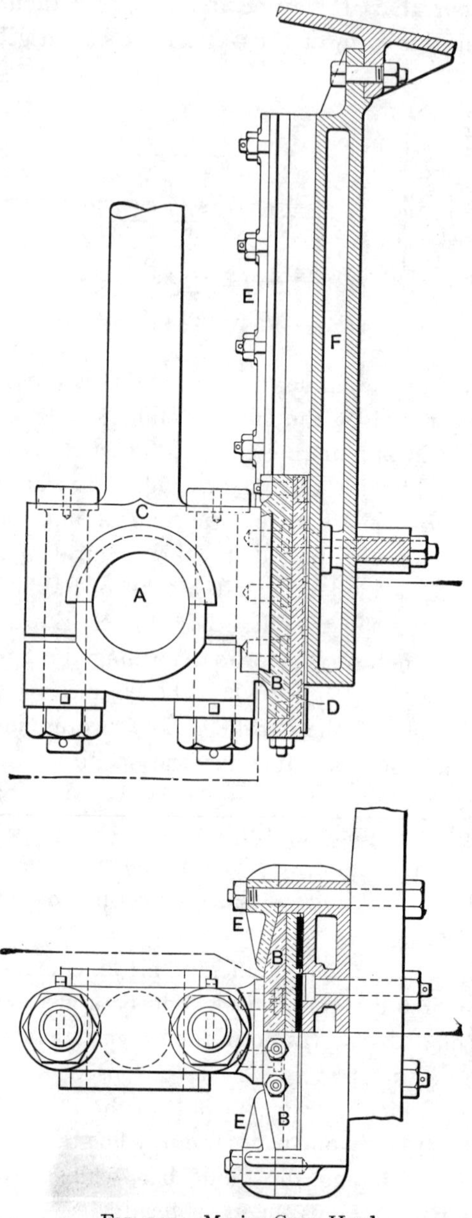

FIG. 141.—Marine Cross Head.

Connecting Rod.—By means of this rod the force from the reciprocating cross head is transmitted to the rotating crank.

It consists of a rod with specially formed ends fitting over the cross head and the crank pins. Figs. 142 and 143 show

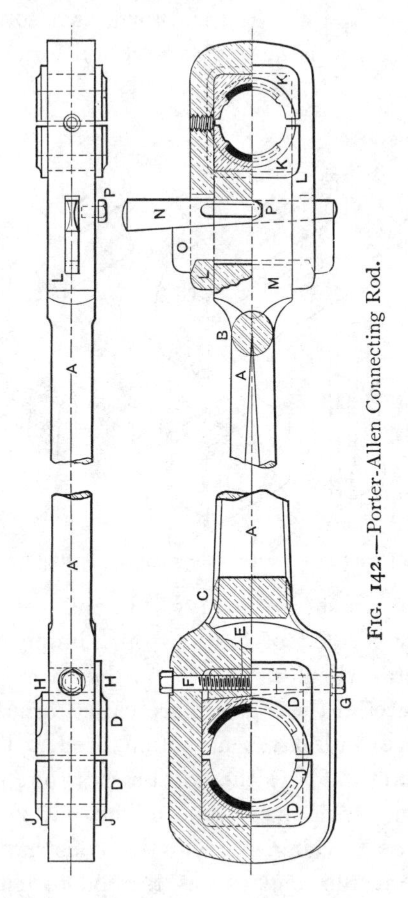

FIG. 142.—Porter-Allen Connecting Rod.

the construction of the Porter-Allen connecting rod. The rod A varies in cross section from the circular neck B at the cross-head end to the rectangle C at the crank end. This enlargement in section is to allow for increase of stress from the neck end, due to the greater oscillation or whipping of the rod as we approach the crank, and to prevent breaking should the rod grip the crank pin. The section of the rod is not truly rectangular at the crank end, as the top and bottom surfaces are arcs of circles. The rod is in reality a frustrum of a cone from which sections have been cut, forming the parallel surfaces.

The crank end of this rod is known as a **solid end** or **box end,** because the rod is here enlarged and formed into a box.

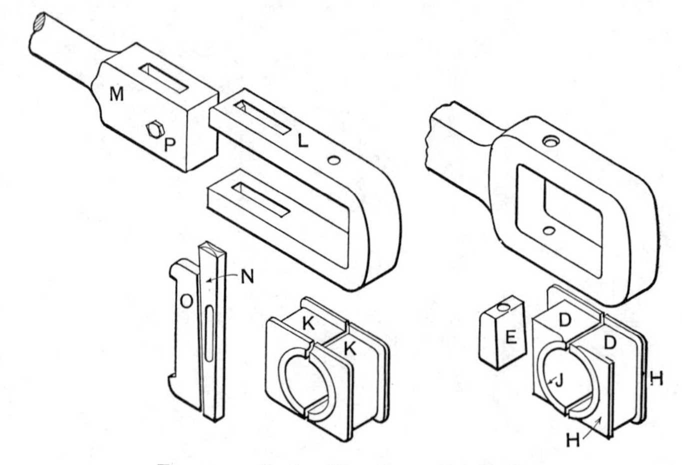

FIG. 143.—Porter-Allen Connecting Rod.

This construction is shown in Figs. 142 and 143. The pieces DD are usually formed of brass with a lining of Babbitt or other white metal where it fits around the crank pin. On account of the material these pieces are called **brasses,** although the names **steps** and **boxes** are often used. The **wedge** E moved by the bolt F forces the two brasses together, thus taking up any wear which may occur. The nut G is a lock nut to prevent F from turning. From the construction the wear on the brasses at this end causes the rod to lengthen. The brasses are held in place by the flanges HH, and the projections or **bosses** JJ on their faces serve to enlarge the surface in contact on the pin, and make a neat fit between shoulders on the crank pin.

The cross-head end, Figs. 142 and 143, is known as a

strap end. In this end the brasses KK are held by the **strap** L. This is joined to an enlarged **stub end** M of the connecting rod by the **cotter** or **key** N and the **gib** O. The key and gib, which pass through slots in the strap and rod are formed as wedges, so that by driving down N the strap is pulled toward the rod, thus tightening the brasses. In this case it is the inner brass which is fixed in position, and, as wear occurs on this brass, the rod becomes shorter. The key N is in contact with the pin end of the slot in the rod, and the gib touches the rod end of the strap slot, as shown in Fig. 142. The key must not touch the end of the slot in the strap, nor the gib the end of that in the rod. When either of these happens it is impossible to tighten the strap further, and a small piece of metal must be placed between the inner brass and the rod or the outer brass and the strap. This moves the strap outward and so allows the key to be used again. These pieces of metal when used for this purpose are called **liners** or **shims.** In some instances two gibs are used with the key between. The key is prevented from working out by the set screw P, and, in some cases, a key replaces the wedge E shown on the box end. The adoption of a strap for one end and a box for the other tends to keep the length of the rod constant.

Marine Connecting Rod.—A box end can only be used when the crank pin projects from the crank disc. When the pin is used between two cranks, a strap end or a modification of it must be used. For large engines the strap is cumbersome, and a type shown in Fig. 144 is employed. This is called a **marine end** or **club end,** and is formed by enlarging the end A of the rod B. To this are bolted the brasses CC and the **cap** D. The brasses are separated by the distance pieces and liners E, so that when the bolts are drawn tight the brasses will not bind on the crank pin. As the brasses wear, the bolts are slacked off, the liners removed, and a piece of soft wire is introduced between the brass and the pin. On tightening the bolts, so that brasses are pulled up, this wire is flattened, giving the amount of play and showing how much must be removed from the liners. The upper end of this rod is forked, and carries the **wrist pin** F, which is shown in the **eyes** GG on the rod. At times the upper forks are made similar to the crank end, and the cross head carries gudgeons fitting into

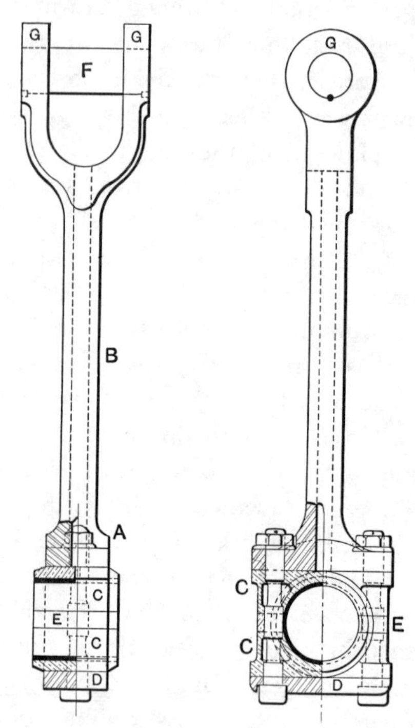

FIG. 144.—Marine Connecting Rod.

them, as shown in Fig. 145. The lower end of this connecting rod is forged to take brasses C fitting around the pin.

Crank Pins, Cranks, and Shafts.—The crank pin is a steel cylinder fastened to the crank or forged solid with it. The most common crank used on the modern engine is the **disc crank,** Fig. 146. This consists of a large **disc** A, usually of cast iron, which is keyed on the shaft B, and to which the crank pin C is attached. A **key** is a small rectangular bar of iron of considerable length, which fitting in corresponding grooves in two bodies prevents their relative motion perpendicular to its length. If the crank is on the end of the shaft it is called an **overhung crank,** to distinguish it from a **center crank,** in which the pin is fastened between two cranks or discs.

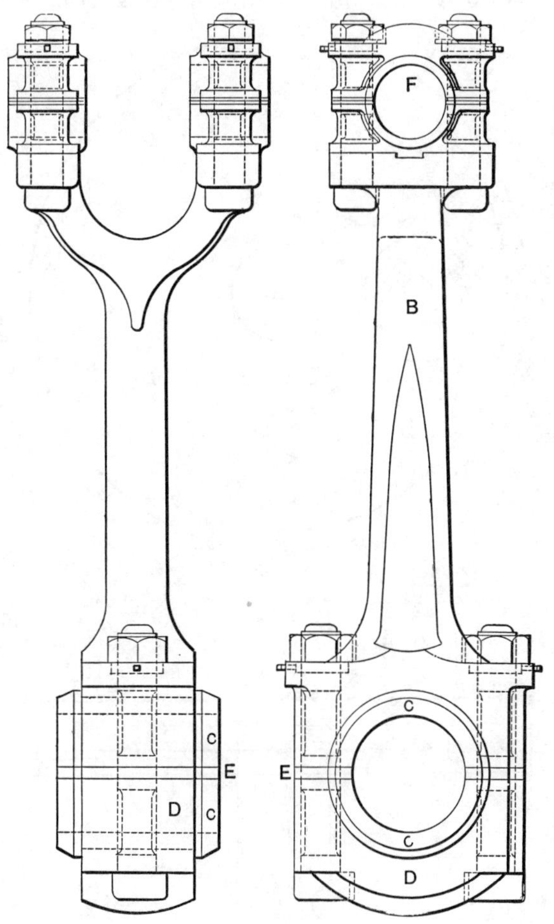

Fig. 145.—Marine Connecting Rod.

These are shown in Figs. 146 and 147. To prevent the connecting rod from leaving the pin, Fig. 146, a plate D is sometimes fastened by the tap bolt E to the end of the pin. The disc is enlarged at F, opposite the crank, to counterbalance the shaking forces produced by the movement of the crank and the **reciprocating parts** (the piston, piston rod, cross head, and connecting rod). The shaft and pin are in many cases forced on the discs by means of an hydraulic press, although at times

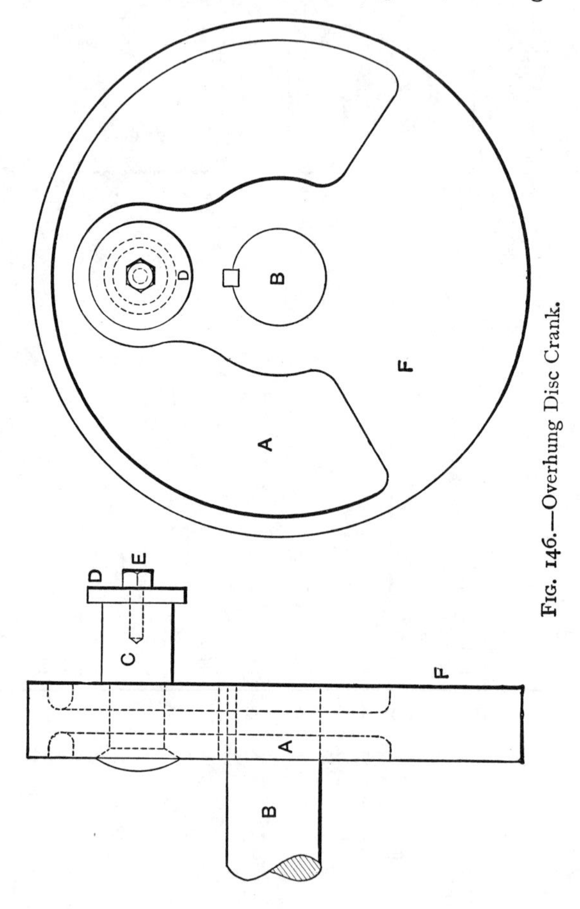

FIG. 146.—Overhung Disc Crank.

this is done by **shrinking**. The latter method may break the disc if care is not taken to have the pin of the proper size. With the hydraulic press the forcing pressure is registered by a gauge, so that, should this become too great, the pin may be removed and reduced in size. A key is used in this case to prevent the disc from turning on the shaft, although the friction

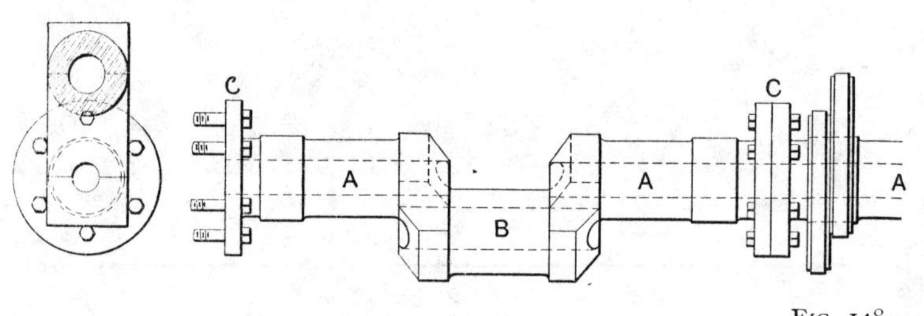

FIG. 148.-

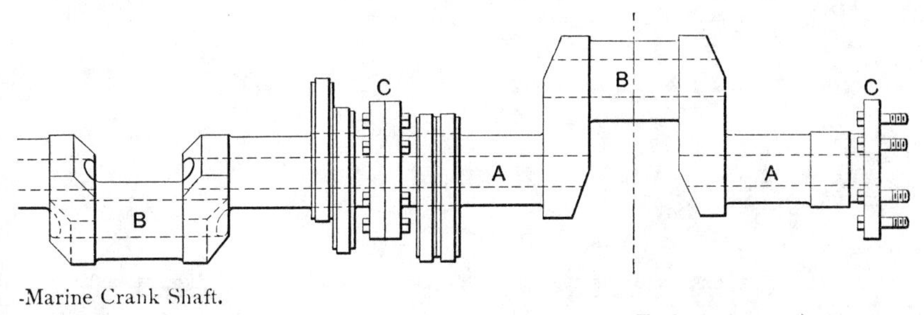

-Marine Crank Shaft.

(*To face page* 139.)

from the pressure produced by forcing is usually sufficient to prevent this.

Fig. 148 shows the crank and crank shaft used on a marine engine. Each section is made of a single forging of high grade steel. As seen from the figure the shaft A and crank pin B

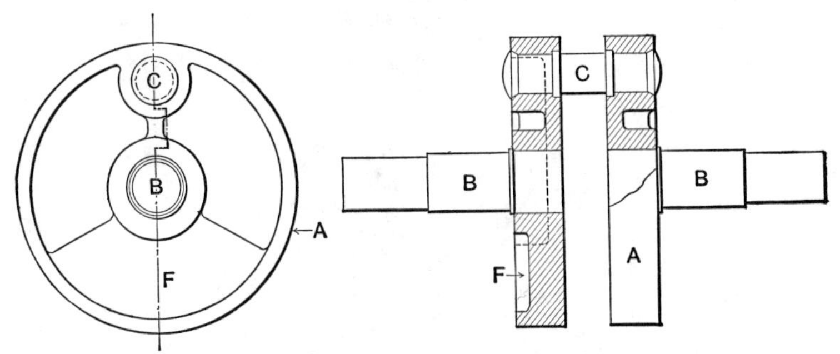

FIG. 147.—Center Disc Crank.

are hollow. This is done to remove the metal from the center where it is apt to be poor, and where it has little structural value, thus reducing the weight of the material. Since several cylinders are placed side by side in marine engines several of these sections are bolted together by the flanges C.

Fig. 149 shows a simple crank arm A, which is often used on slow-speed engines. The pin C and shaft B are keyed as

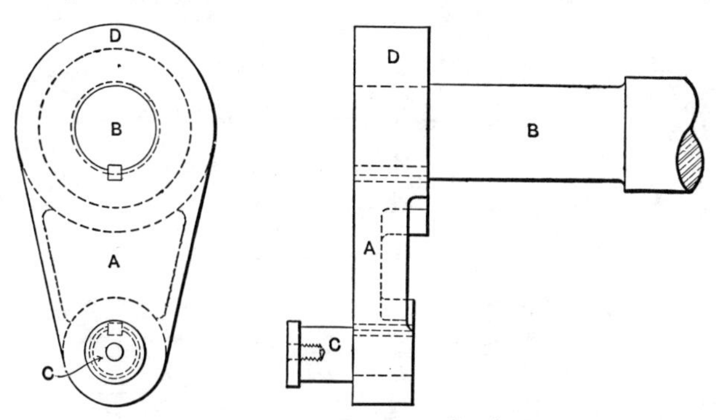

FIG. 149.—Overhung Crank.

shown. The boss D, or enlargement at the center, is to give additional strength.

Shafts.—The shafts of engines are made of high-grade steel. When small they are made solid, and for the large sizes hollow. These shafts are made as large as 25 inches in diameter.

A marine engine shaft is shown in Fig. 148, and a stationary center-crank engine shaft in Fig. 147.

Fig. 150 shows the form of shaft used on American locomotives, where the crank pins D are attached to the driving

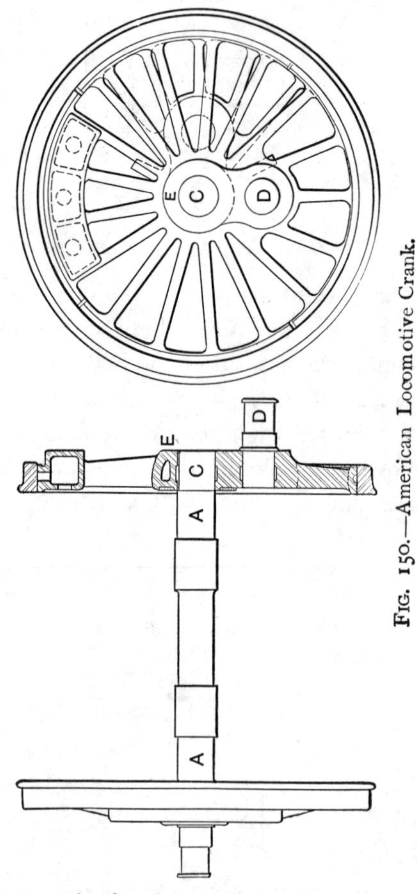

FIG. 150.—American Locomotive Crank.

wheels. A, termed the **journal,** is the part which fits the bearing, and the portion C is forced into the **wheel center** E. As shown in Fig. 151, English locomotives are often made with inside cranks. It will be noticed that the cranks are "**quartering,**" or set at right angles, so that when one crank is on dead center the other is near the point at which it can receive the maximum thrust. This is necessary in the case

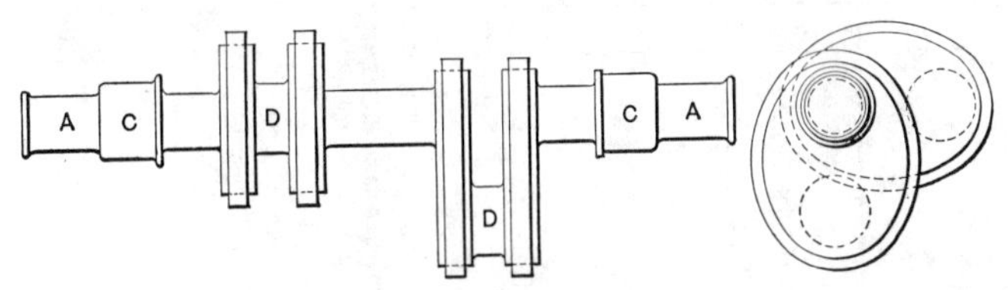

FIG. 151.—English Locomotive Crank.

of all engines which are to be stopped and started frequently or which are to start under load.

Bearings.—The shaft of an engine is supported by castings forming part of the frame of the engine, or is bolted to it or to a separate foundation. The castings, Fig. 152, known as a **bearing,** contain boxes or brasses A, C, D, similar to those used in the connecting rod, which are held down on the main casting by a cap B. The journal rests in these boxes. They are made of cast iron fitted with Babbitt metal and are sometimes divided into four parts that the wear may be taken up by the **side or cheek boxes** CC and the **top box** D. When the wear on the **lower box** A becomes great, a liner is placed beneath, bringing the center of the shaft again to the center of the engine. The bearings may be made in two parts only, and, in such cases, the line of division should be perpendicular to the direction of maximum thrust. The bearing on the engine frame is termed the **main bearing,** while that on the **pedestal** outside of the frame is the **outboard** or **pedestal bearing.** The portion of the shaft which turns in a bearing is called the journal.

Locomotive Bearings.—The bearings for a locomotive axle, termed **journal boxes,** are made so that they may slide within the frame, bearing against a **spindle** which carries the load from the engine frame. This is done through springs, which are so arranged in conjunction with levers that the pressure on each bearing is a certain portion of the weight of the locomotive. The box, Fig. 153, is of cast iron with a brass or bronze bush A inserted. As the pressure is always downward

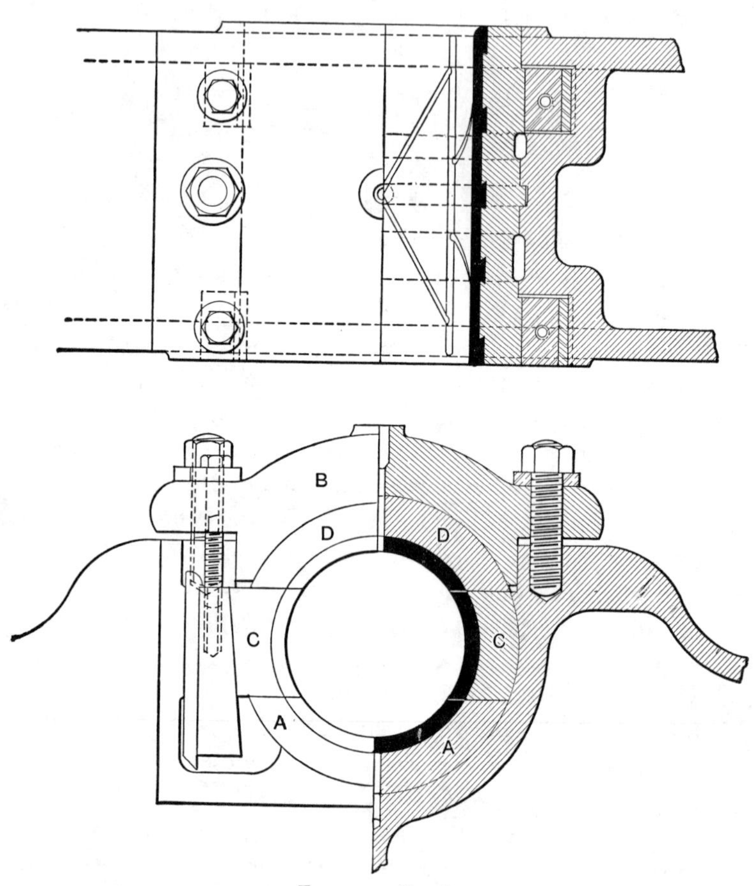

FIG. 152.—Bearing.

this is the only part lined. The cellar B is used to hold oily waste and prevents the journal from leaving the surface of the box. The projections CC fit on either side of the engine frame and hold the box in place.

Thrust Bearing.—On steamships and in certain other instances where there is pressure on the shaft in an axial direc-

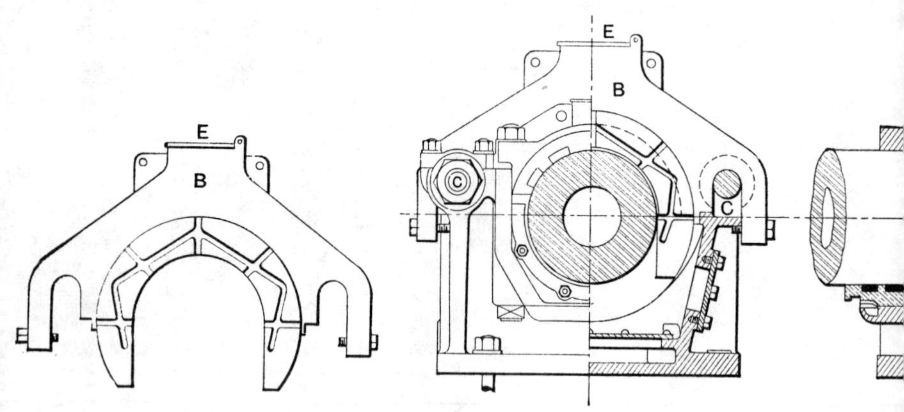

Fig. 154.

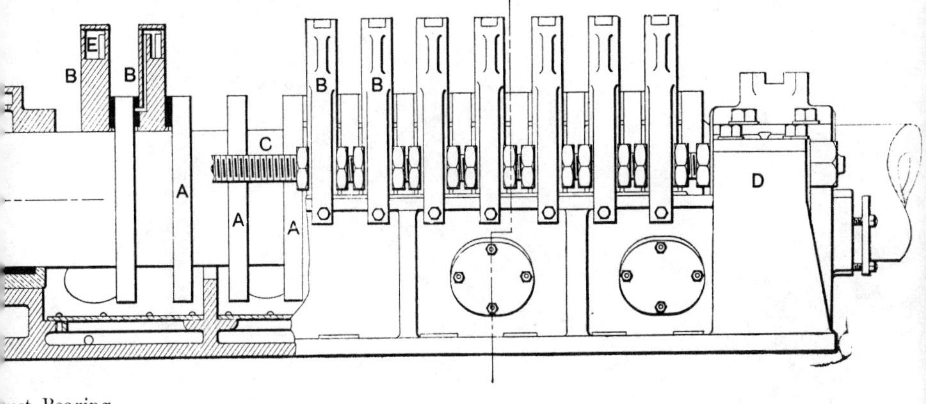

·ust Bearing.

(*To face page* 143.)

tion some means of preventing motion being produced by that pressure must be provided. This is accomplished by means of a thrust bearing, as shown in Fig. 154, which represents the thrust block of a large marine engine. The collars A on the

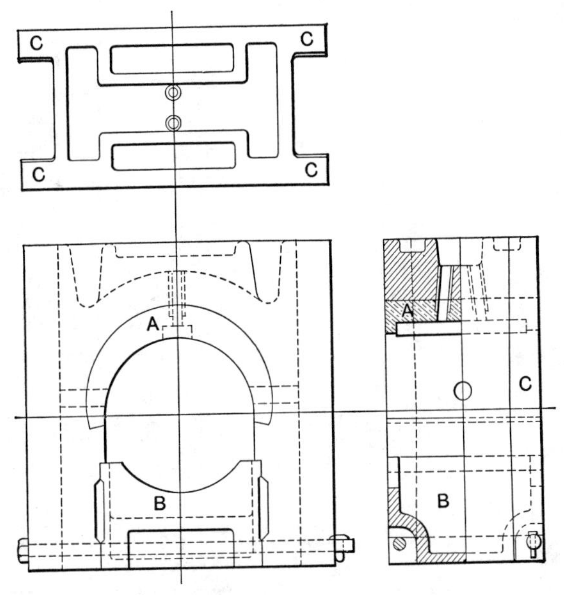

FIG. 153.—Locomotive Bearing.

shaft fit between the thrust yokes B, which transmit the thrust through the bolts C to the main housing D, which is bolted to the frames of the ship.

The separate rings are lubricated from oil boxes E, the oil being fed down by strands of wicking syphoning the oil from the box into the oil pipe. The bottom of the rings on the shaft run in a mixture of water and oil, the bottom of the inside of the bearing being made water tight for the purpose.

A step bearing, such as is used for a vertical shaft, is shown in Fig. 155.

The shaft A fits into a depression in the cast-iron shoe B, which is caused to rotate with the shaft by the key C. The shoe B has a small circular projection in the center, which fits into a similar depression in the brass plate or disc D, and a projection on the lower side of the plate fits in a depression in the block below. The sides of this block, forming a square, fit between four abutments F, and prevent the block from rotating with the shaft. The bottom of the block is spherical, which

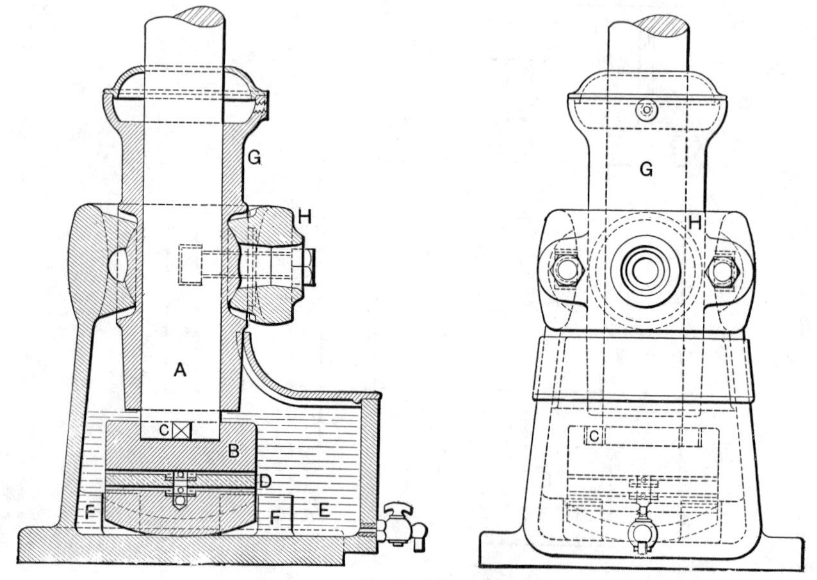

FIG. 155.—Vertical Step Bearing.

allows the upper face to come true against the brass plate. These pieces are placed in the bottom of an oil reservoir E. Oil entering the block from E, flows to the center and rises through a small hole to the brass plate and shoe.

There are two semicircular grooves in the bottom of the shoe and plate, and as these revolve the grooves cause the oil to lubricate the surfaces.

Above the **step** is the bearing G, held beneath the cap H, keeping the shaft in line, and, by the ball and socket joint at the center, permitting of adjustment.

Fly Wheels.—The construction of a solid fly wheel is shown in Fig. 156. This form is only used for small wheels. The **hub** A of the wheel is keyed to the shaft, and is joined to the rim B by the **arms** C. This rim is turned off to better balance the wheel, and to form a smooth driving surface for the

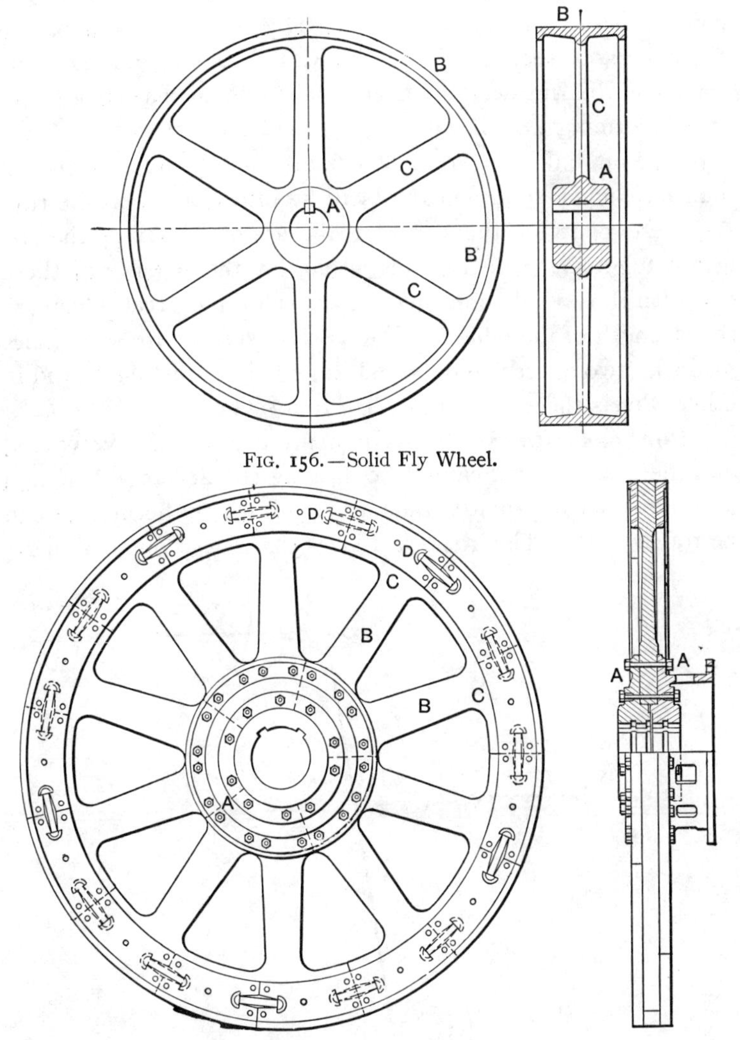

FIG. 156.—Solid Fly Wheel.

FIG. 157.—Sectional Fly Wheel.

belt. If the wheel is not balanced in this way, it is necessary to add small weights to perfect the balance, to prevent irregular stresses being set up by the unbalanced wheel. For large engines, fly wheels are built in sections, an example being shown in Fig. 157. The **spiders, or discs,** AA are forced on the shaft and keyed; and to them are bolted the arms B, each pair of which carry a section of the rim C. The extremities of the arms and the rim sections C are planed off so that they fit together, forming a solid whole. To hold the rim sections C together, iron links are used which are first heated, so that in their expanded condition they will fit into slots D on the rim. As these contract they grip in these slots, holding the rim firmly together. Steel plates fitted on the outside of these cast-iron rims break joint on opposite sides and greatly increase the strength of the wheel. The central spiders are sometimes made in two pieces and bolted together on the shaft, and in other wheels there is a single arm to a section.

Rockers and Slides.—In many engines the valve rod and the eccentric rod cannot be directly coupled together, and it is necessary to provide some means by which the motion can be transferred. The rocker, Fig. 158, is a lever or arm A,

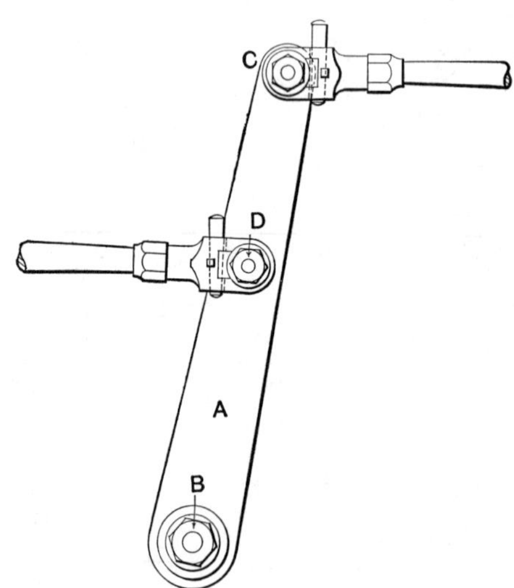

FIG. 158.—Corliss Rocker Arm.

mounted on a shaft B. To the other end of this lever, an arm or pin C is attached. The bearings for the rock shaft B are

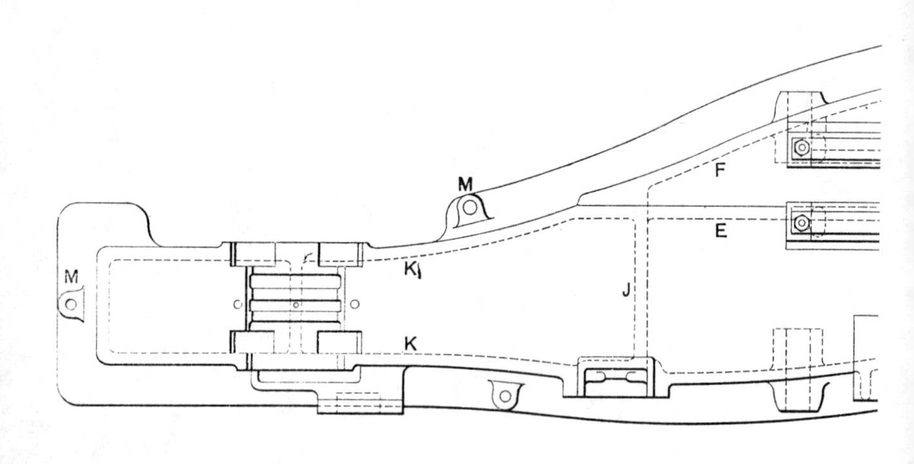

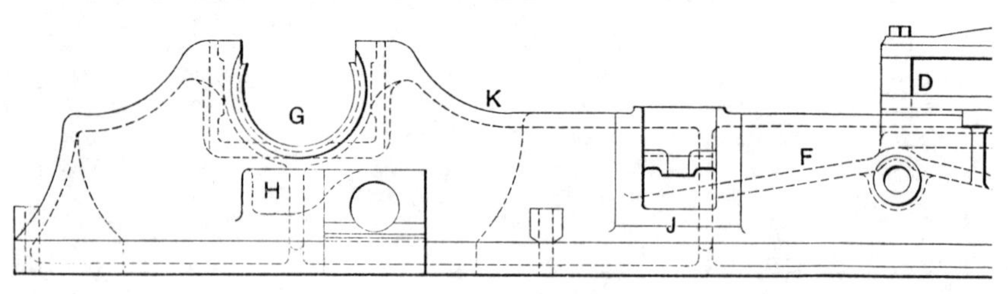

Fig. 160.—Porter-

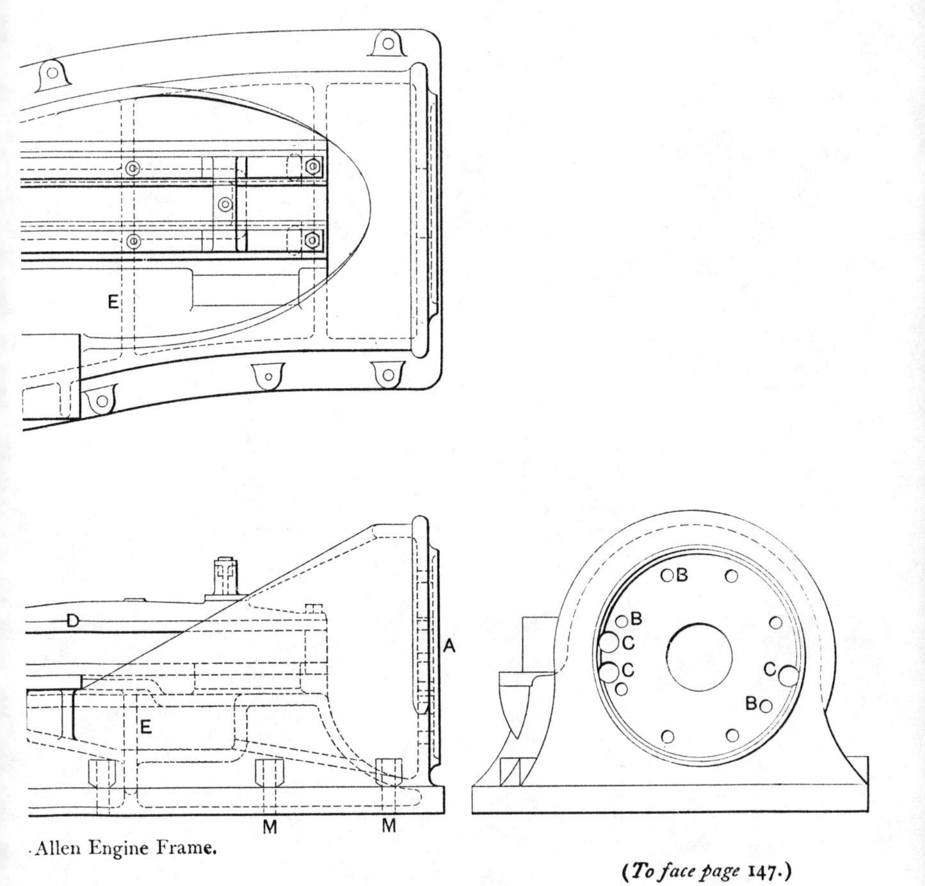

·Allen Engine Frame.

(*To face page* 147.)

bolted to the engine frame, so that the pin D is in the plane of the eccentric rod and C in that of the valve rod. A slider is shown in Fig. 159, and is used in the engine shown in Fig. 110. In this

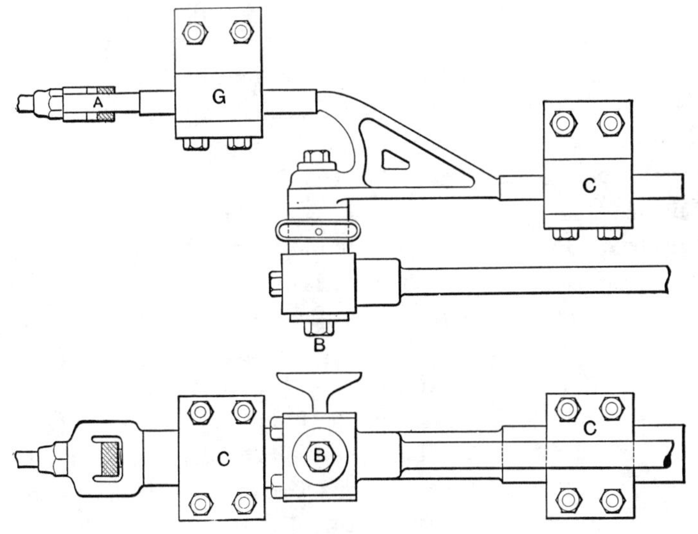

FIG. 159.—Weston Valve Slider.

case the valve rod is attached at A, and the eccentric rod turns on the pin B. This is the equivalent of a cross head, except that the reciprocating rod and the oscillating rod are not in one plane. The guides C and G constrain the slide to move correctly.

Engine Frames.—In Figs. 160, 161, 162, 163, and 164 are found examples of engine frames. The object of the frame is to hold the cylinder, piston rod, and shaft of an engine in proper alignment. To do this it must be made stiff, and yet as light as possible, and the metal must be properly distributed for this purpose.

Fig. 160 shows the frame or housing of the Porter-Allen engine, sometimes called the **Tangye** frame. The cylinder overhangs from the end A, being bolted to it by stud bolts from the cylinder, which extend through the holes BB. The valve stems extend through the larger holes CC. Two of the four guides DD are formed in the frame, which is provided with the webs EE beneath these parts to stiffen it. The surface F, as shown in the side view, is inclined, to allow for the movement of the connecting rod. G is the opening for the main bearing, into which the boxes are fitted, as shown in Fig. 152. Since the thrust of the piston rod is not in a line through the center of this box, the couple, produced by the force on the piston, tends to twist the frame, and, to prevent this twisting, the webs H and J are added, and the sides and top K of the main casting are made heavy. This frame is bolted to a masonry foundation by large **foundation bolts,** which extend through the holes M. The **foundations** are masses of brick or stone masonry or concrete, which are of sufficient size to properly distribute the weight of the engine and to prevent vibration.

Fig. 161 is an example of the **bed plate** or frame for a Corliss engine. In this type of frame, a heavy casting N is bolted to the foundation by the bolts extending through the holes M. This casting contains the main bearing G and a yoke X, to which the **guide barrel** O is bolted. This heavy yoke and barrel form of guide is adapted to do away with the springing apart of the guides which occurred with the earlier forms of Corliss frame. The outer end A of the barrel is bolted to the cylinder, which is also bolted to the foundation. The most recent form of frame designed by the Pennsylvania Iron Works Company is shown in Fig. 162. In this the guide barrel is carried by the main frame, which prevents the guide from deflecting and makes a stiffer frame. The cylinder, which is bolted to A, rests on the foundation but is not bolted to it, as is done in the case shown in Fig. 161. Figs. 161 and 162 represent modern practice for large horizontal engines. The webs which are shown are added for stiffness. The crank revolves in the space R which has a bottom cast in it to catch the oil which comes from the guides and crank pin, as will be explained later. The projection P is used to support the rocker for the eccentric. The guides in the barrel O are bored out for a cross head similar to the one shown in Fig. 139.

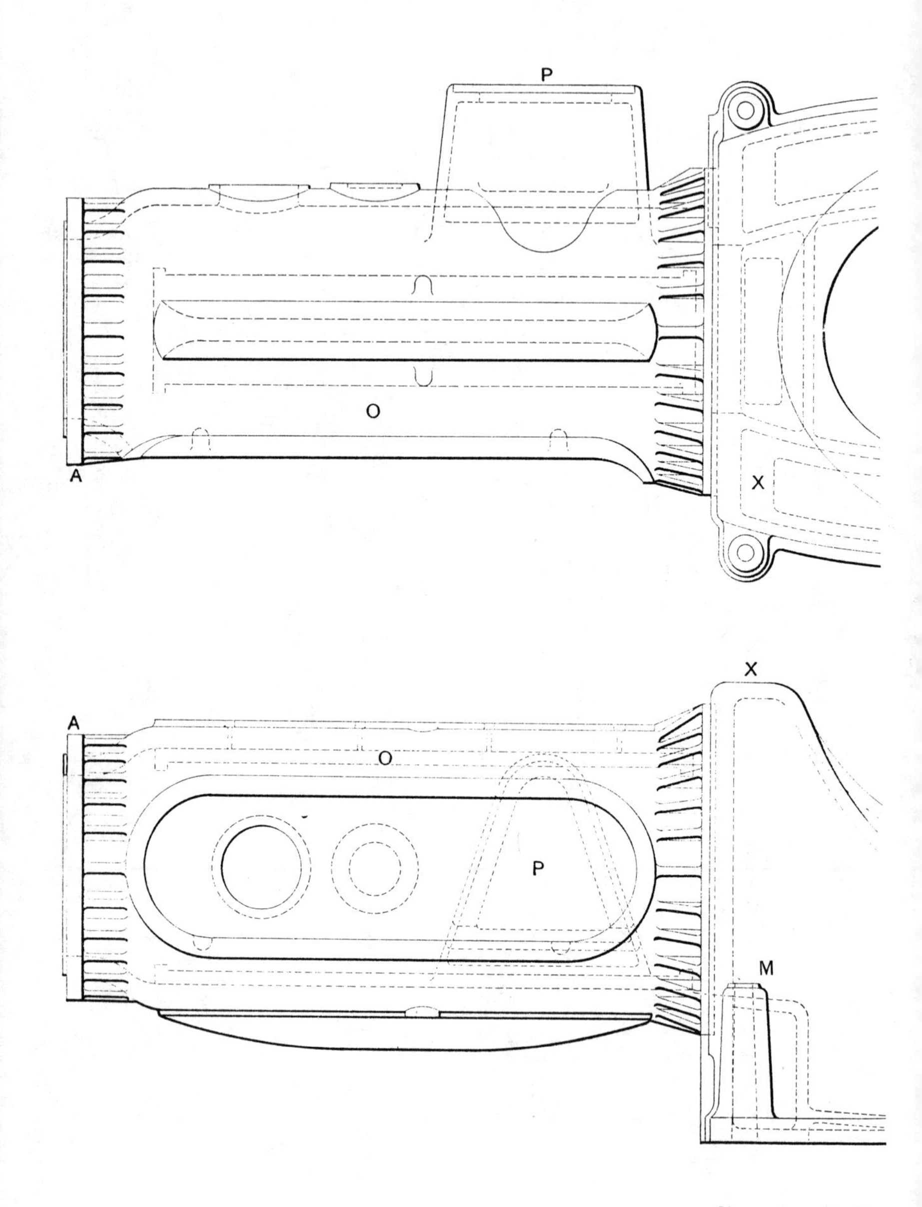

FIG. 161.—Corliss

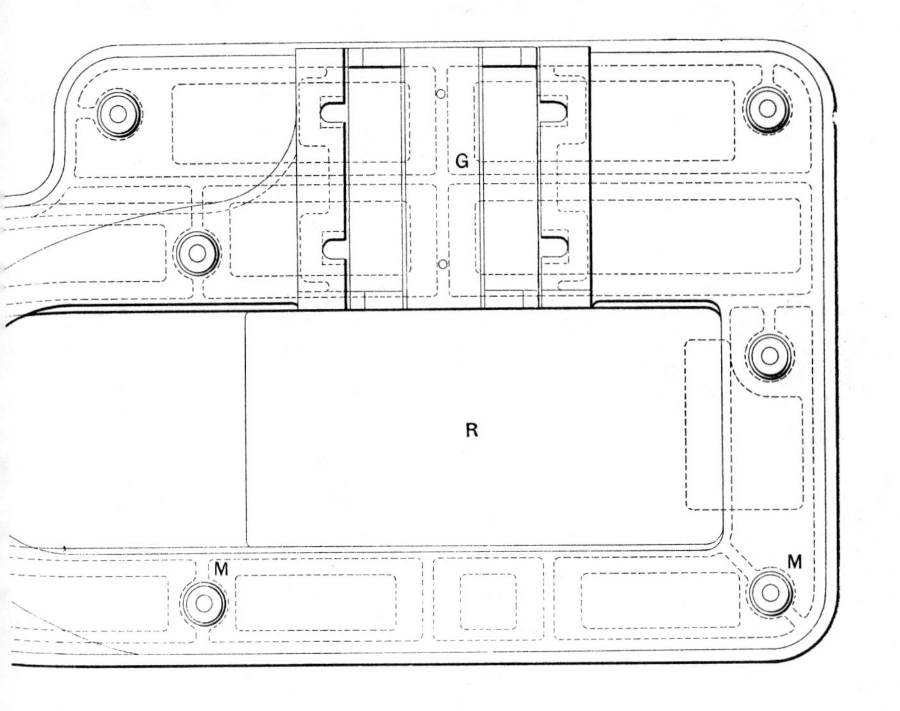

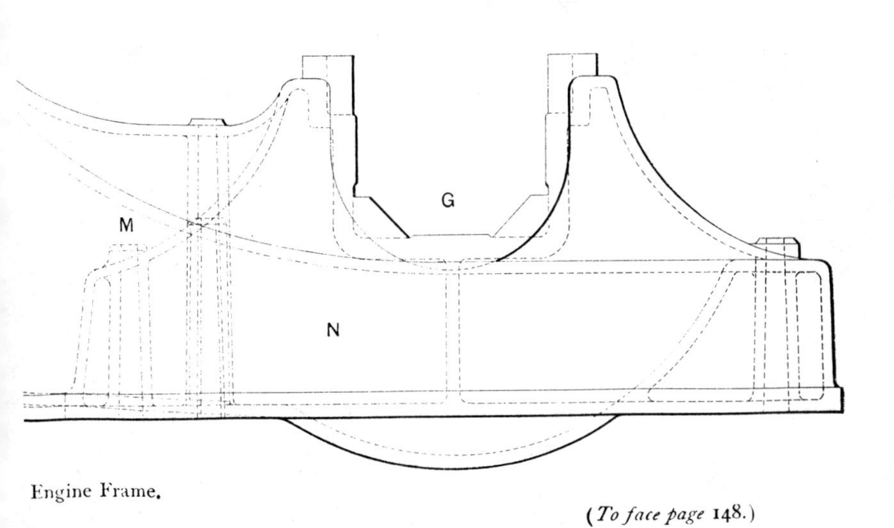

Engine Frame.

(*To face page* 148.)

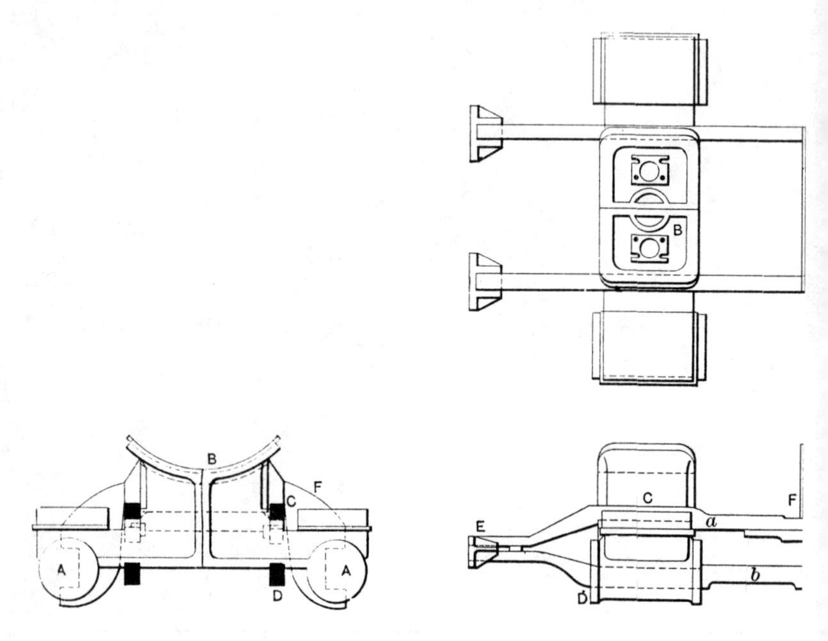

FIG. 163.-

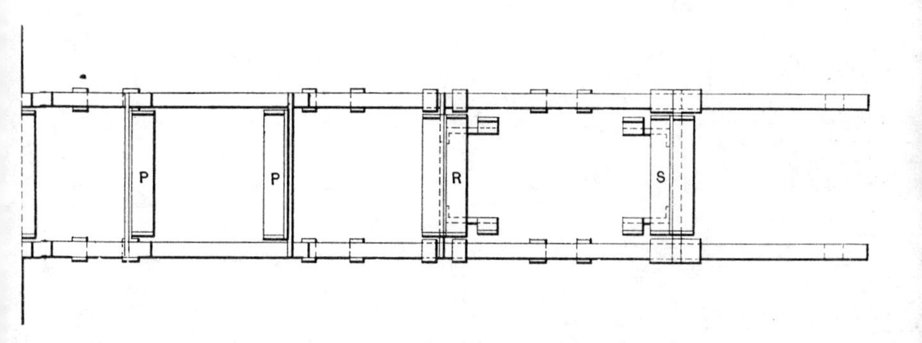

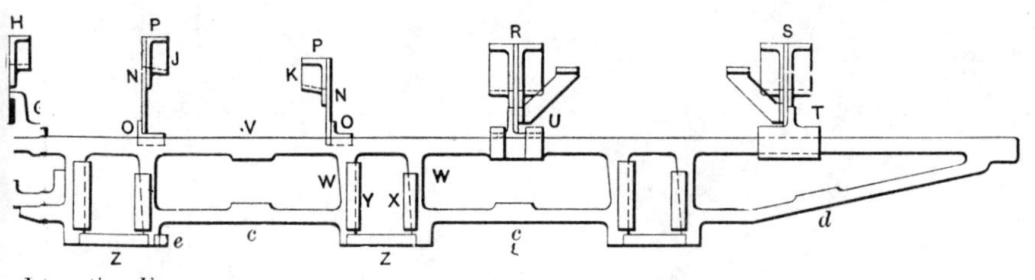

–Locomotive Frame.

(*To face page* 149.)

Locomotive Frame.—The cylinder castings A, Fig. 163, are bolted together, forming the saddle B, to which the smoke box of the boiler is bolted, as shown in Fig. 12. This saddle contains the passages leading the steam to and from the cylinders. The frame is bolted to the cylinder casting at C and D, passing above and below it. At the front E is the frame-filling piece, and to this the bumper is bolted. The blackened portion at F is a section of the guide beam which is fastened to the

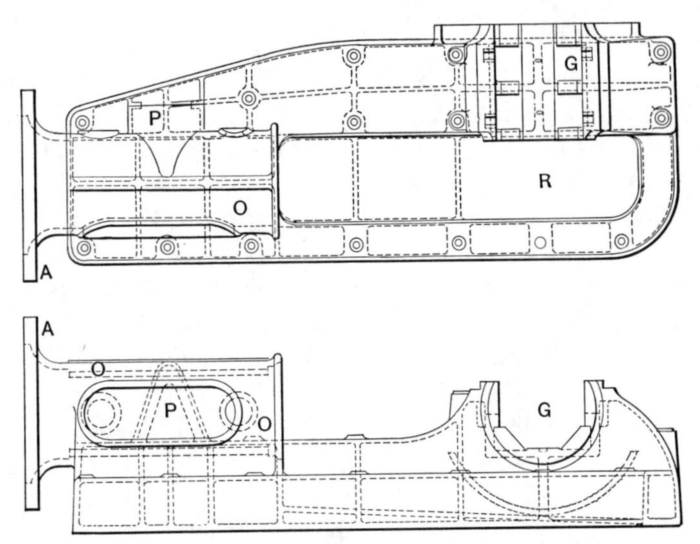

FIG. 162.—Corliss Engine Frame.

knees G and extends beyond the frame to carry the guide bars. The projections HJK are **bearers,** which consist of plates N, connected at the bottom to a tie forging O uniting the frames, and at the top to the angle iron P, which is attached to the shell. The bearers at R and S are attached to the shell but not to the frame, the joint on the frame being a sliding one, the forging either extending over the side as at T, or beneath straps as at U. The frame shown here is intended for the Vanderbilt type of fire box. With the other types, the fire box either rests on the frames at the back or extends between them, being carried by an **expansion pad** bolted to the side. This is nothing but an angle iron bolted to the side of the fire box which rests on the frame, and so carries the weight, but allows for the expansion. The portion V is called the **top rail,** while WW are the **pedestal legs,** one of which is finished on an angle. This with the **pedestal wedge** X and the **pedestal gib** Y allows for adjustment to compensate for the wear on the sides of the journal box, which is shown in Fig. 153. The wedge X is attached to a bolt which passes through the **pedestal cap** Z, and permits the adjustment of the wedge.

The portions *a* and *b* are respectively the **top front rail** and the **bottom front rail ;** *c* and *c* are the **middle braces,** and *d* is the **back brace ;** *e* is a cross tie extending between the frames.

Marine engine frame.—The marine frame, Fig. 164, consists of a bedplate A, bolted to the frames of the ship, to which are attached the front **columns** BB, and the back **A-frames** CC, supporting the cylinders. These A-frames are sometimes used back and front, and in the more modern engines steel columns are used throughout. This latter is true when it is especially desired to reduce the weight. The A-frames shown in the figure carry the guides DD, which are so constructed as to envelop the cross-head, as shown in Fig. 141.

The bedplate contains the main bearings EE. The sections of this casting are so made as to facilitate manufacture and erection. The tie rods FF brace the structure and prevent vibration. In the back view of the engine, the upper end of the inside tie rods and the lugs G, to which they are attached, and which are part of the cylinder casting, are omitted. In the column construction the guides are attached to cross braces extending between the back columns. In the construction of vertical engines for stationary practice two A-frames are usually used, although the latest practice is to cast the front and rear frame below the guides in one piece, and to surmount this by a barrel guide, the lower portion being bell shaped, with a portion cut away over the main bearing.

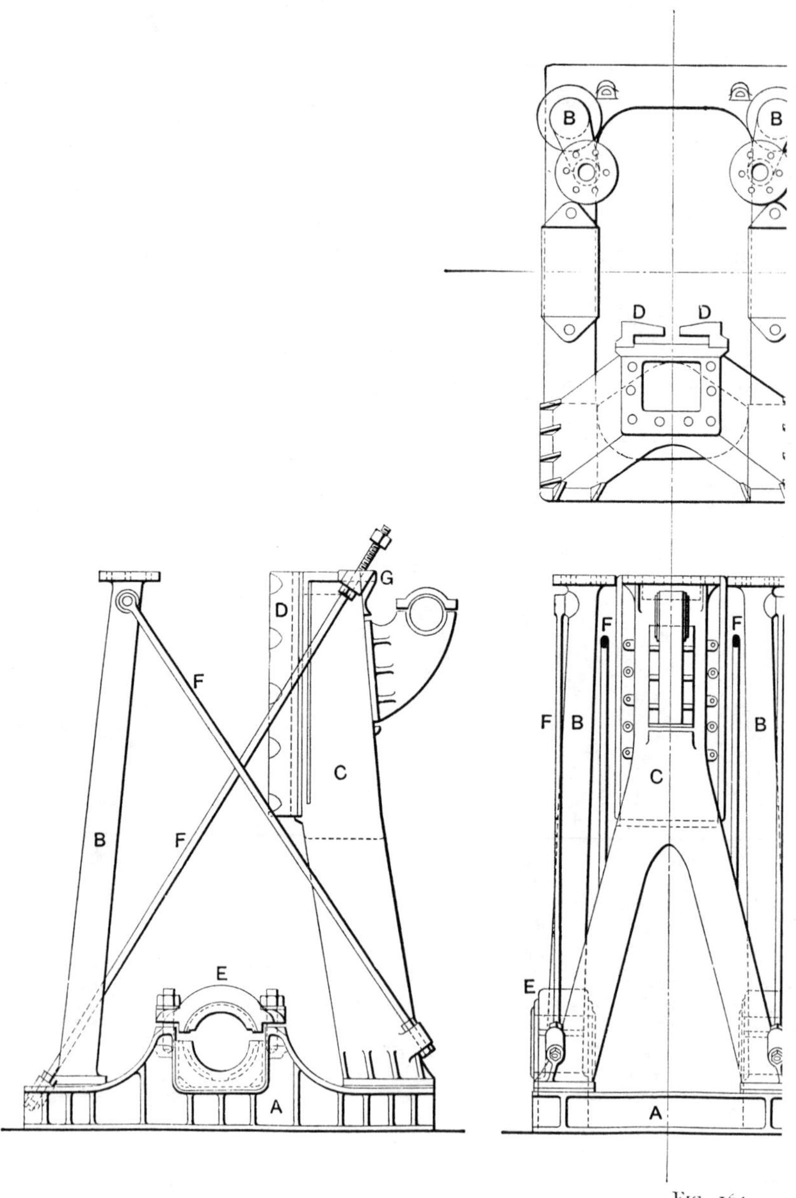

FIG. 164.

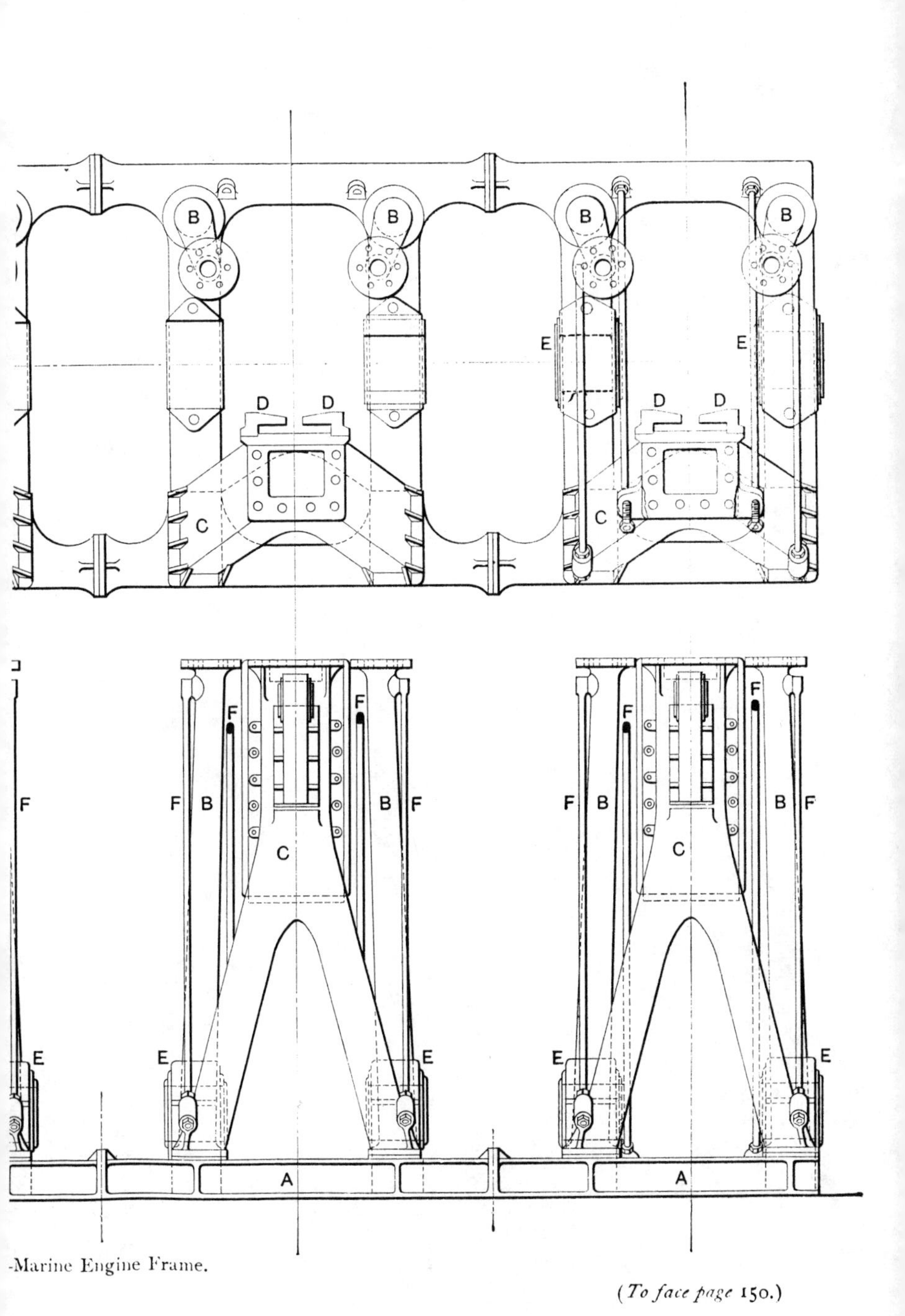

-Marine Engine Frame.

(*To face page* 150.)

Lubrication.—To reduce the loss of power due to the friction of the various sliding surfaces, lubricants of various kinds are used. The most common lubricant is mineral oil, which is introduced between the surfaces and forms a film, the particles of which roll over each other and permit of easy movement. The friction depends apparently more on the lubricant and manner of lubrication than the materials of the surfaces. This lubricant is fed to the different parts through lubricators.

Sight-feed lubricators. — A **Sight-feed lubricator** is shown in Fig. 165. It is used to introduce heavy cylinder oil

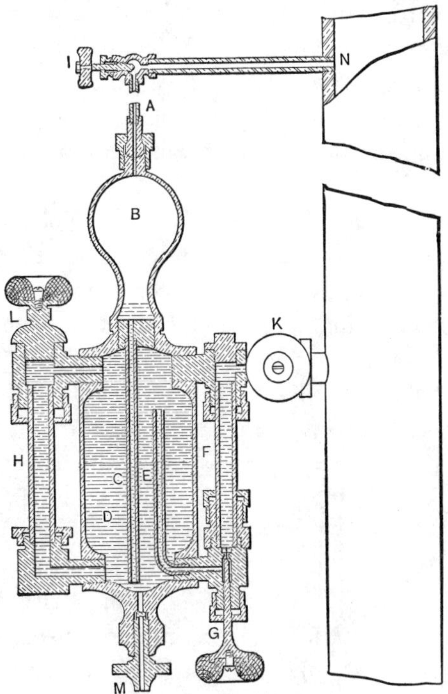

FIG. 165.—Sight-feed Lubricator.

into the steam entering the engine. Steam in the pipe A and in the **condensing chamber** B condenses and falls through the tube C to the bottom of **reservoir** D, from which it displaces the oil. This oil is forced through tube E, and ascends in drops through the water in the gauge glass F, being regulated by the **needle valve** G. The gauge glass H shows the amount of oil in the reservoir. When necessary to fill the lubricator the valves I and K are closed, and, after removing the cap L, the water can be drained out through the waste cock M. The reservoir is then filled through the opening at L, and, after replacing the cap, the valves are again opened. The steam pressures at the points N and K are practically the same, and, for this reason, there would be no tendency for the oil to flow, due to the steam pressure, but as water collects in the chamber B and the pipe above it, the pressure produced, forces the water downward displacing the oil. A very slight difference in level between the points N and K is all that is necessary and in cases where no difference in level can be obtained an inverted "U"-tube is used, the condensation in the pipe-leg falling back into the pipe while that in the lubricator-leg causes the flow of oil.

The oil used in a cylinder should be of such a nature that it will not decompose nor vaporize under the high temperature, but, if it does decompose, harmless compounds only should be formed. Oils of animal and vegetable origin form acids which attack the iron of the engine and boiler and should not be used. The most satisfactory oils are the heavy mineral oils remaining after distilling the light oils from crude petroleum. These oils will stand the heat and give satisfaction.

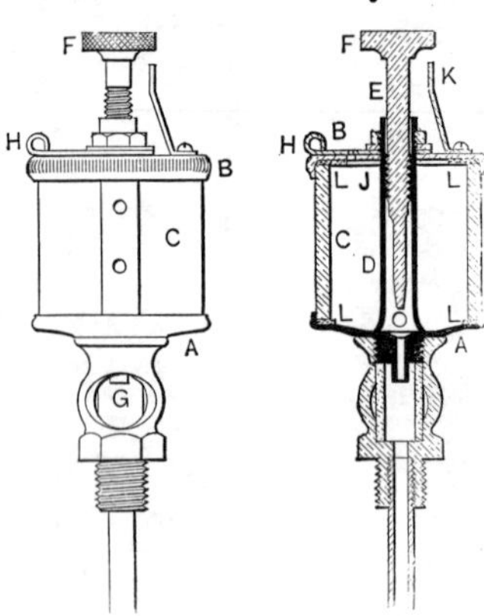

FIG. 166.—Sight-feed Oil-cup.

Sight-feed oil-cups.—For oiling the bearings with light mineral oil sight-feed oil-cups, Fig. 166, are used. These are made with a brass base A, and cap B, between which is a glass barrel C. Through the hollow post D passes the spindle of a needle valve E, which is controlled by the knurled head F at the top. By adjusting the head F, the spindle, which is threaded, lifts the valve from its seat and regulates the supply of oil flowing through the holes at the bottom of the post. This oil can be observed as it drops through the glass tube G in the base. The cup is filled by turning the sliding-cover H from the hole J in the cap. The spring arm K fits against a flattened place on the knurled edge and prevents the head from turning, thus making the flow uniform. The joint at L is made on a washer of cork.

Lubricators for the Pins.—The lubrication of the crank pin is accomplished in several ways. With an overhung crank the arrangement shown in Fig. 167 is most often employed. Oil from the cup drops into the spout A, which projects into a cavity in the ball B. The hollow arm C, carrying the ball B, is screwed into the cap D on the crank pin E, and is of such a length that the ball is opposite the center of the shaft. The cap D is bolted to the end of the crank pin and prevented from turning by the **dowel** G. When the oil enters this cavity in B from the spout A, it is carried by centrifugal force to the outer end, and thence through a passage F made in the crank pin to the pin surface. The standard H is attached to the engine frame. Another method which is used more often with inside cranks and on cross-head pins is to have a wiper A, shown in Fig. 168, move below a wiper tip B or a stretched piece of wick. In this way the drops of oil are wiped off and fed to the rubbing surface.

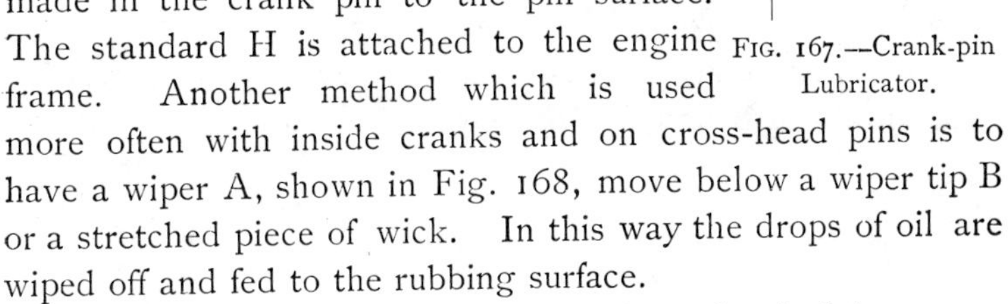

FIG. 167.—Crank-pin Lubricator.

A center crank is sometimes oiled from the shaft by means of small holes made in the crank, through which the oil is forced by centrifugal force.

The cross head is sometimes lubricated by means of lubricating grease in addition to an oil feed. This grease is a heavy product of mineral oil, which is plastic at ordinary temperature, but which acts as a good lubricant when the temperature rises slightly. A **grease cup** is shown in Fig. 169. In this the grease is driven out by the pressure exerted on the piston A by the spring B, the nut C and the cap D not being in contact.

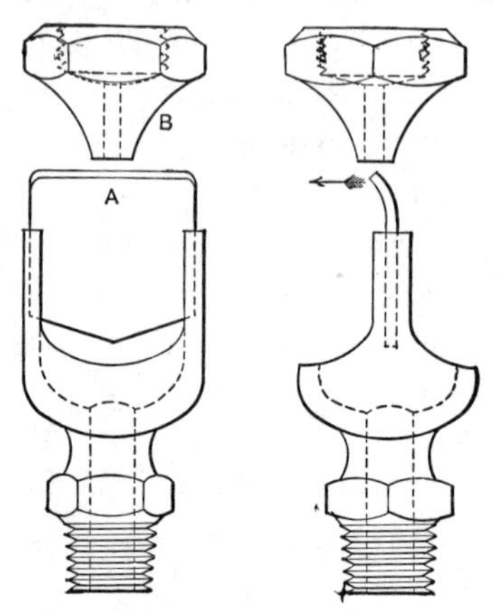

FIG. 168.—Cross-head Lubricator.

The rate of feed is regulated by the screw at E. When it is desired to stop the supply of grease the **thumb nut** C is screwed down, lifting the piston and relieving the pressure.

For cylinder lubrication, where high-pressure steam is used, powdered graphite has been proposed, because of the liability of the oil to volatilize. Fig. 170 shows the form of a Lunkenheimer sight-feed **graphite lubricator.** Graphite is introduced through A, and the amount of feed is regulated by B, steam entering through the valve C. D is the drain plug through which the condensed steam can be drawn off. The glass face F permits the observations of the feed. The cock E closes the discharge.

Hand-pump cylinder lubricators.—This form of lubricator, which is used where cylinder lubricators are thought unreliable, or to supplement them, is shown in Fig. 171. It consists of a brass base A, in which is the ball-valve B and the piston C. The piston has a hole with a conical base through its center. The pump rod is not rigidly attached to the piston, the piston being free to move a slight amount on the rod. The enlarged end of the rod on striking the small bushing E at the top of the piston raises it. When the piston is pushed downward the conical end of the piston rod is forced into the conical hole at the bottom of the piston. The oil reservoir F

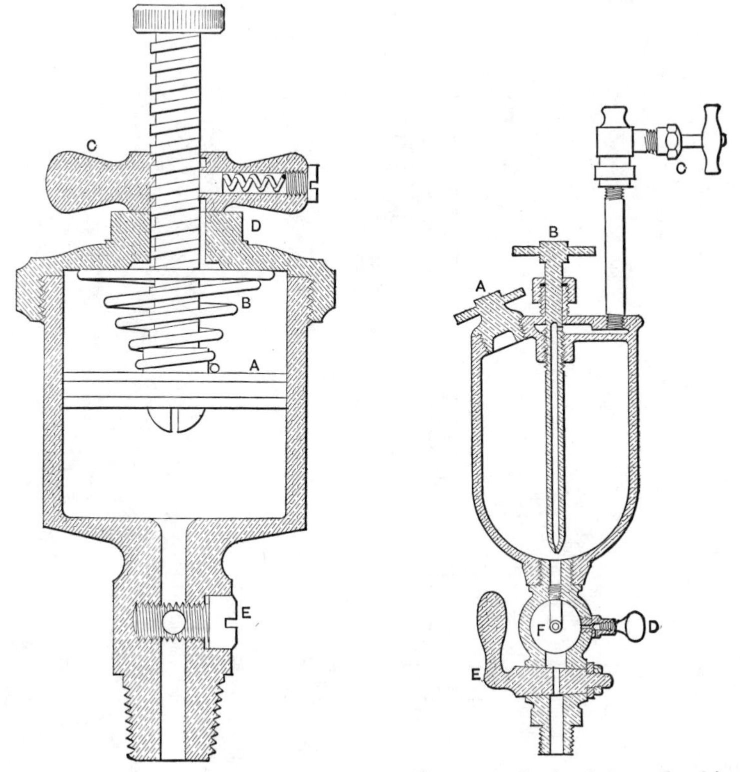

FIG. 169.—Grease Cup.

FIG. 170.—Lunkenheimer **Graphite** Lubricator.

is made of glass. On raising the handle G, the piston is raised and oil is sucked below the piston, flowing through the flutings H on the piston rod. The ball-valve B at this time is held up by the steam pressure below. As the handle is pushed down the rod is forced against its conical seat, and the oil is forced into the steam pipe passing the ball valve.

Positive lubricator.—To render cylinder lubrication continuous and positive, the Sterling lubricator, shown in Fig. 172, may be employed. This consists of an oil reservoir A,

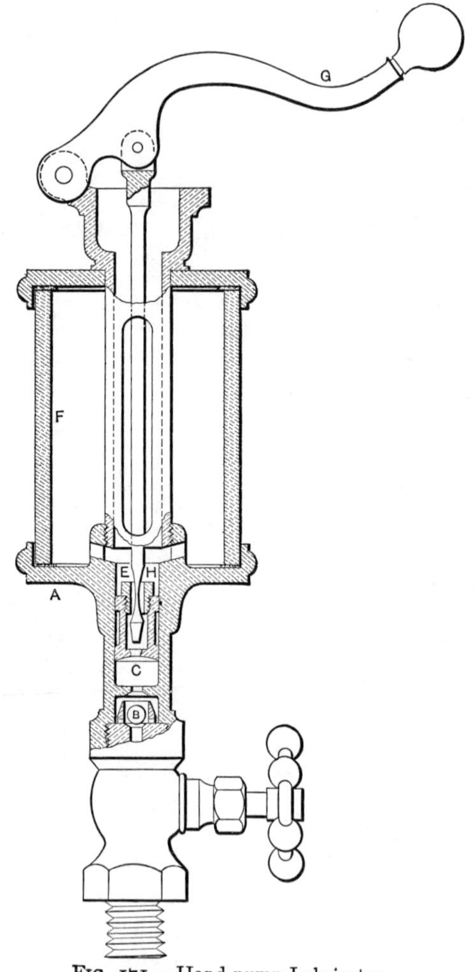

FIG. 171.—Hand-pump Lubricator.

carrying at its base two pump cylinders. The plungers B of these pumps are driven by connecting rods, which are connected to an oscillating beam C. This beam is supported midway be-

K K G K K G H STERLING LUBRICATOR COMPANY ROCHESTER N.Y. PATENTS PENDING

FIG. 172.—Sterling Positive Lubricator.

(*To face page* 157.)

tween the points of attachment of the connecting rods, and is caused to oscillate by the crank D. The crank is carried on a shaft E, extending through the oil reservoir. This shaft is driven by ratchet wheel F, which is moved by the rod G, attached to some portion of the engine having a reciprocating motion.

The amounts of motion of the pawl arm H, and of the connecting rods are adjustable. These, together with the adjustment of the throw of the crank, makes it possible to greatly vary the amount of oil supplied. The pawl-arm adjustment regulates the number of strokes, and the crank and connecting-rod adjustments regulate the stroke. The discharge from the pumps is through the three way cocks KK, which are so constructed that the discharge can be allowed to flow from the

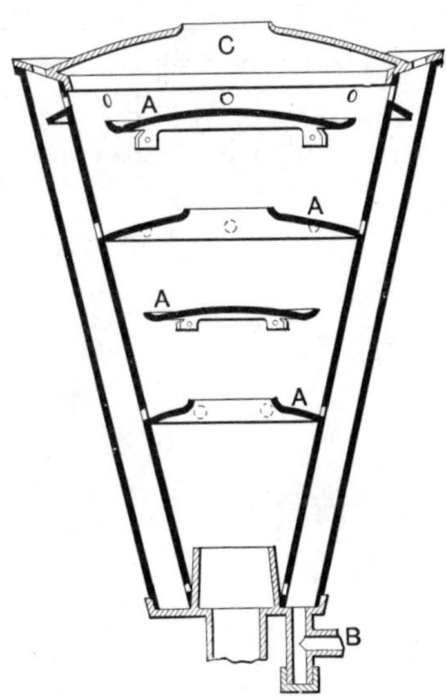

FIG. 173.—Exhaust Head.

spouts shown in the figure that the amount of the feed may be determined. The sight glasses on the side shows the amount of oil present in the reservoir.

It has been found advisable to run oil in a continuous stream on the bearings, pins, and guides of large engines. The oil is then collected, filtered, and pumped to reservoirs, from which it flows by gravity. In some cases the oil is introduced under pressure to these parts, and in other instances the crank, connecting rod, and cross head are inclosed, so that the rod dipping into oil and water at the bottom of the case splashes the mixture over the pins and guides. In one engine this action also forces the oil into the main bearing. These continuous systems have been found to give excellent results in the reduction of friction, wear, and cost of oil.

Exhaust head.—To prevent the discharge of oil and water from the exhaust pipe of an engine it is customary to place an exhaust head at the top end of the line. This head, Fig. 173, is constructed with a series of baffle plates AA, which are so arranged that the water and oil are collected and discharged through the pipe B, the dry steam discharging from C at the top. The head is really a separator. The discharge of oil and water from an unprotected pipe is not only a nuisance in disfiguring property, but it is apt to cause decay in the iron roof on which it falls.

CHAPTER VI.

VALVE MOTIONS AND DIAGRAMS.

THE common valve motion of to-day is that produced by an eccentric, directly attached to the valve rod or combined with one or more levers to accomplish the desired results. One form of the eccentric is shown in Fig. 174. It is a circular disc of

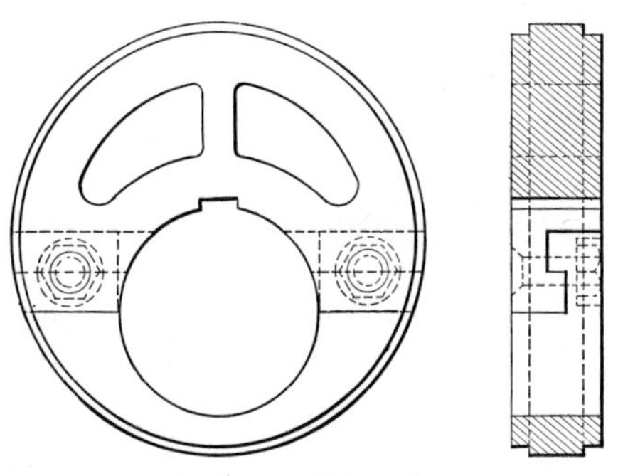

FIG. 174.—Eccentric.

metal which is fitted to a shaft, but in which the shaft does not pass through the center of the disc. The rotation of this disc produces a reciprocating motion at the end of a connecting rod attached to the eccentric. The disc is usually called the **eccentric** or **sheave,** and the connecting rod is called the **eccentric rod.** The eccentric is really a crank of small throw, but one with the crank pin enveloping the shaft. Consider the short-throw crank shown solid in Fig. 175. If the crank pin shown by the small solid circle be enlarged until it becomes the large dotted circle, we have the eccentric as shown in Fig. 174. Hence, whenever the eccentric is used it may be treated as a crank, the throw of which equals the distance between the cener of the disc and the center of the shaft. The eccentric is seldom used except where a small amount of motion is desired, and on account of the great friction with a short throw and large disc, the eccentric is used only to give a reciprocating motion from a shaft and not to change a reciprocating motion into rotary motion. As a shaft is weakened by the introduction of a center crank, and as these are expensive to

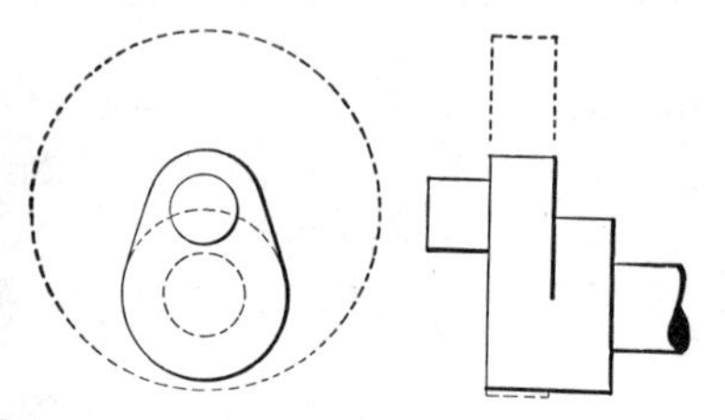

FIG. 175.—Eccentric and Equivalent Crank.

make, the eccentric has been introduced to produce the reciprocating motion for auxiliary parts situated between bearings. In such cases the plain shaft is strong and easily finished, and the eccentric readily attached after construction. If the motion required can be obtained from the end of a shaft, a simple crank of short throw would cause less friction loss than an eccentric, and is often used in such cases.

The eccentric strap A, Fig. 176, is required to attach the eccentric and rod. The **strap** A is made in two parts, which are bolted together with the bolts BB. This is free to turn on the eccentric, but is prevented from moving axially by the groove C in the strap, which fits over a projection on the eccentric. To this strap the eccentric rod is bolted at D, which rod is attached to a valve rod. The eccentric is lubricated by an oil cup attached at E, the oil from which gradually discharges. The eccentric sheave is sometimes fastened to the shaft by means of **set screws** or keys, and in some cases it is forged with the shaft. A set screw, Fig. 177, has its end cupped and so hardened that when it is screwed tightly to the shaft it enters the metal and prevents the parts moving relatively to each other. A key is used as shown with the sheave of the eccentric, Figs. 174 and 176. Fig. 176 shows another method of bolting the halves of a sheave together.

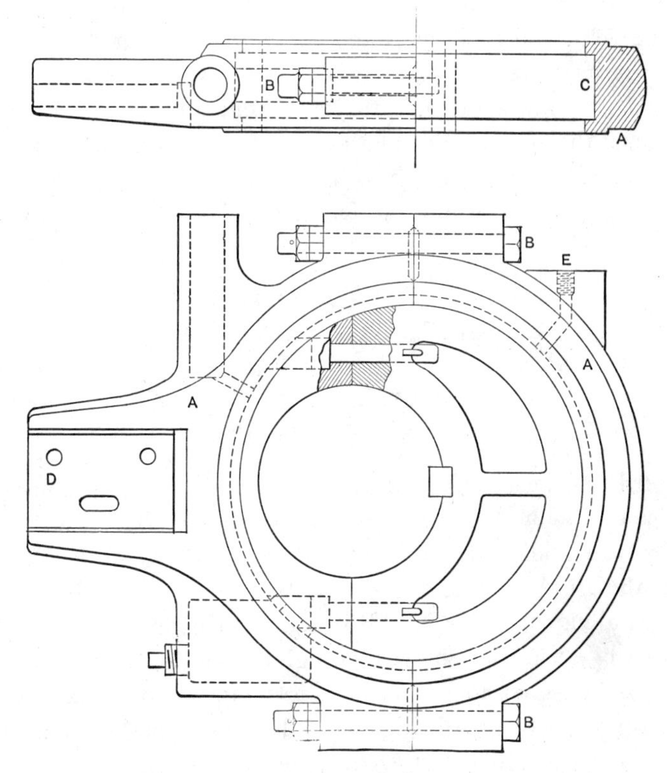

FIG. 176.—Eccentric Strap.

When the eccentric rod is bent to clear an obstruction, its action is the same as that of one which, in the plane of the sheave, directly joins the center of the sheave to the center line of the pin on the valve rod or slide.

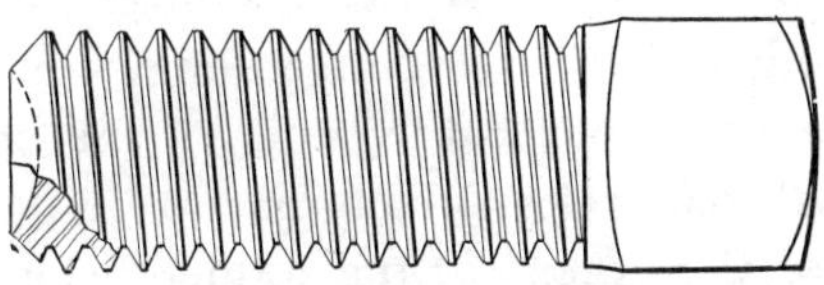

FIG. 177.—Set Screw.

The action of the eccentric and rod can be considered as that of the crank and connecting rod represented in Fig. 178.

OA is the eccentric and AB the rod. As AO rotates about O the point B is moved back and forth in the straight line BO. When A is in the position A′ the point B is at its extreme left hand position, and when at A″, it is at the extreme right of its travel.

As A moves from A′ to A the end of the rod B would move, if the rod remained parallel to the center line of the engine, to B_2 along the arc $B'B_2$. The horizontal movement of B would then be B′C, which is equal to A′D. If now the

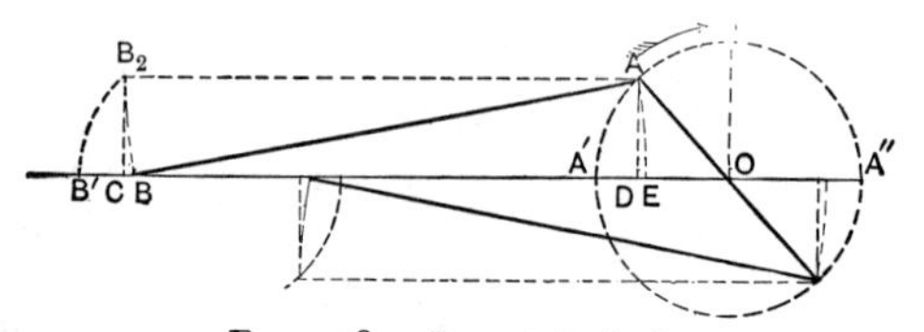

FIG. 178.—Eccentric Action.

point B_2 be swung from B_2 to the center line of the engine, about the point A, it will move along the arc B_2B, which produces the additional horizontal motion CB, which is equal to DE, the arc AE being drawn with B as center. As B actually moves along the line OB′, B′B is the actual horizontal motion of B, while the crank moves through the angle A′OA. This is seen to be equal to A′E, which is composed of the two parts A′D and DE, the first being due to the motion of the crank AD, and the second to the movement of the rod AB. This latter is said to be the effect of the **angularity** of the rod. On the return-stroke the effect of the angularity of the rod is to decrease the motion of the rod rather than increase it, as is shown on the lower side of Fig. 178. It is seen that the longer this rod the smaller the value of the distance DE, and for this reason the motion of the piston of an engine is quite different from that of a valve, since the ratio of the connecting rod to the crank throw is less than that of the eccentric rod to the throw of the eccentric, the former being seldom over 6 to 1 while the latter is often 20 to 1. The effect of angularity is great in case of the piston motion, but so small in the case of the valve that we may consider its motion as if due to the eccentric radius alone, in which case the motion is the same on both strokes. Under these conditions, then, the motion of the valve from its extreme position may be considered as equal to the distance AD, and the valve is in its middle position when the crank is in the position OA‴, and A′O is the movement of the valve from its extreme position. The position of the valve half-way between the two extremes is termed the **mid position,** and it is customary to measure the movement of the valve from this point. The movement of the valve from its middle position, then, is found by projecting the eccentric radius, in any position, on a line representing the eccentric when the valve is at one extreme position. Thus in Fig. 178, DO gives the movement of the valve from its mid position. The valve controls the steam to the cylinder, admitting it or exhausting it at the proper time. It may have various forms and be actuated by different means, but to study the action of the eccentric and valve it is well to consider the simple D **slide valve** which is shown in Fig. 179. This valve is a casting containing a cavity

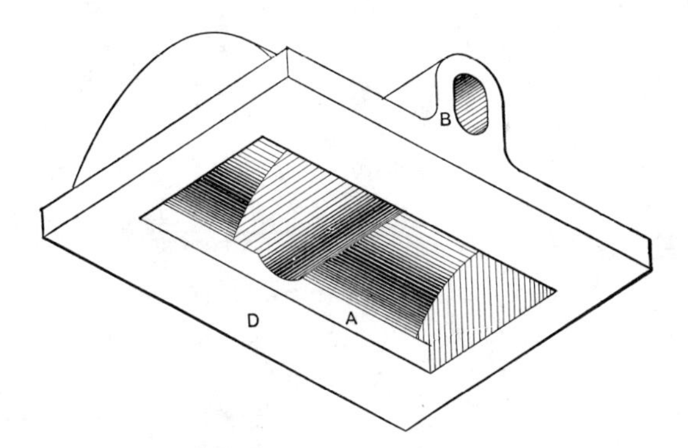

FIG. 179.—D Slide Valve.

A, as shown in the figure. The opening shown at B passes through the valve body and is intended to receive the valve rod which drives it. Modifications of this valve are extensively used on locomotives and high speed engines.

Fig. 180 shows the cross section through a valve and passages when the former is in its mid position. The surface C, on which the valve moves, is called the **valve seat,** and the lower surface of the valve is the **valve face** D. The openings in the valve seat are called **ports.** These ports are the ends of the **steam passages** EF and the **exhaust passage** G. Between these passages are the **bridges** HH.

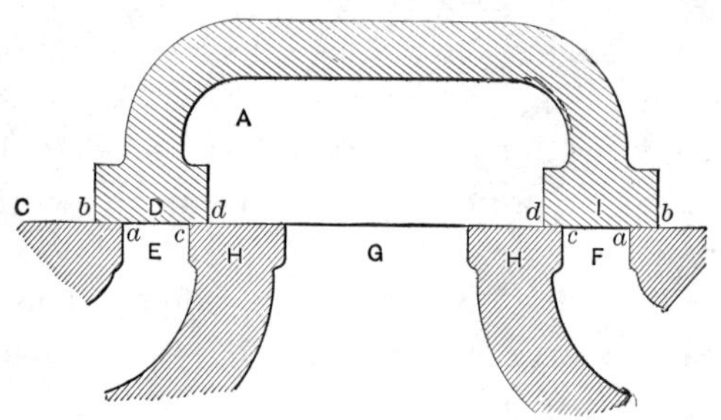

FIG. 180—Steam Valve and Passages.

Let the relative position of the valve and piston and of the crank and eccentric be as shown in Fig. 181, the valve being made just large enough to cover the ports instead of as in Fig. 180. If the engine be turned in a counter-clockwise direction from the position shown in A, Fig. 181, the eccentric will move the valve so that steam is admitted to the right-hand side of the piston, and should there be any steam on the left of the piston it can escape through the left steam passage under the valve and through the exhaust port to the air or a condenser. As the steam continues to move the piston, the eccentric moves the valve further to the left, opening the port wider until the **maximum port opening** occurs, when the crank has moved through 90 degrees from the head center, as at B, Fig. 181; that is, from the position of the crank in which the piston is at the head end of the cylinder. In the circles D, E, and F the heavy lines are intended to show the relative positions of the crank and the eccentric. The further movement of the piston then moves the eccentric back again, making both the admission and the exhaust openings smaller. The valve again reaches its mid position when the engine is on its crank center at C, Fig. 181. As the fly wheel continues to move the shaft the valve moves further to the right, opening the left-hand end of the cylinder to the steam, and the right-hand to the exhaust, the opening of the exhaust being called

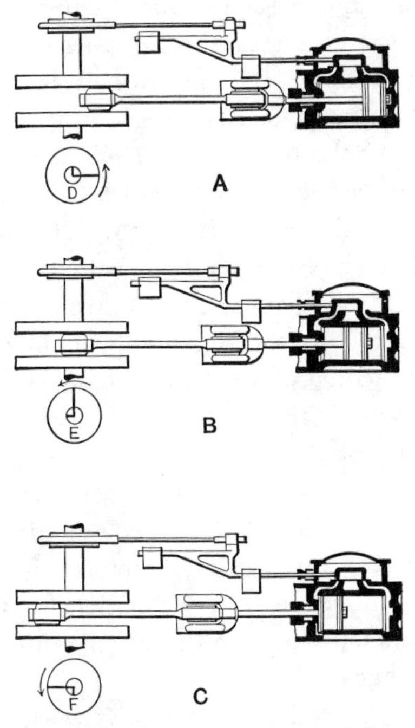

Fig. 181.—Action of a Square Valve.

the release. When the crank reaches the vertical position opposite E in the figure the valve has been moved as far as possible to the right and begins to close again. When the engine reaches the head center, the valve is again in its mid position, and both steam and exhaust ports are closed, the engine being as shown at A, Fig. 181.

To have the steam port open at the beginning of the stroke the eccentric is moved on the shaft so that the angle between the crank and eccentric in Fig. 181 is more than a right angle. This allows steam to be admitted before the beginning of the stroke, and exhaust to take place before the stroke is completed. The amount of port opening at the beginning of the stroke is called the **lead**. It is noticed that with the valve shown in Fig. 181 the exhaust on one side occurs at the same time that admission takes place on the other, and also that as soon as the admission ceases exhaust begins, so that we have no chance to expand the steam. For these reasons we extend the valve, so that in the mid position it laps over the steam port on each side, as shown in Fig. 180. The amount the valve extends over the port on the steam side when the valve is in its mid position is called the **steam lap,** while that on the exhaust side is called the **exhaust lap.** It is important to remember that these distances are measured when the valve is in a position midway between its extreme positions. The amount *ab*, Fig. 180, is the steam lap, and *cd* is the exhaust lap. These are called **outside** and **inside laps** respectively, from their positions, without reference to whether they are on the steam side or not.

To give the valve lead when lap is used, the eccentric must be moved still more. The angle through which the eccentric is moved to shift the valve from its mid position to that occupied at the beginning of the stroke, or, what is the same thing, to move it an amount equal to the lap plus the lead, is called the **angle of advance.**

In Fig. 182 the action of the valve with lap and lead is shown. When the piston is at the beginning A of its stroke, steam has been admitted on one side and the exhaust port opened on the other. As the piston moves, the port opening becomes larger, and in this instance the maximum opening occurs earlier on account of the angle of advance. Then in a short while at B cut off occurs on the right and the steam expands. Before the exhaust port is opened on the head side of the piston that on the crank side has been closed, and compression of the steam has commenced, as at C. By the time admission takes place on the crank side this compression has raised the steam to a pressure much above that of the exhaust.

A shows the position of the parts at the beginning of the stroke; B shows the valve cutting off steam on the head end; C

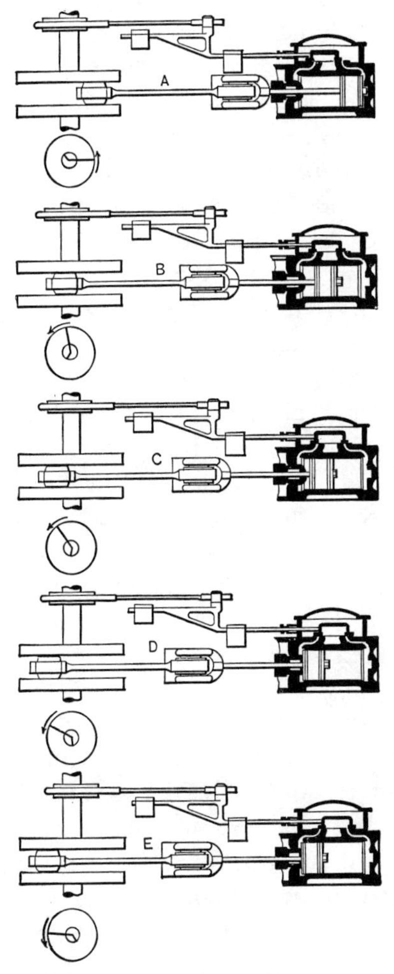

FIG. 182.—Valve with Lap and Lead.

shows the exhaust port closing on the crank end, and D shows the exhaust for the head end opening. E shows admission taking place on the crank end. Similar figures might be drawn for the return stroke. The port is wide open for only the instant that the valve is reversing its motion. It may be necessary sometimes to have this action continued for a greater period, and for this purpose overtravel is used. **Overtravel** is the amount by which the valve uncovers the bridge by moving beyond the edge of the port As this is accomplished by the use of a larger eccentric the movement of the valve is quicker, and consequently the cut off is sharper.

Valve Diagrams.—As has been noted before, the motion of the valve from its middle position is equal to the projection of the eccentric radius on a line corresponding to the position of the eccentric when the valve is in its extreme position. This is true with a long eccentric rod and when the valve has the same motion as the end of the eccentric rod. In some cases the motion of this latter point is modified by a system of levers, so that the motion of the valve is a multiple of that of the eccentric-rod end. In this case a direct connection may be considered to exist between the valve and an eccentric of larger size, without the lever being used.

Suppose OA″, Fig. 183, represents the actual position of the eccentric, and OC the position of the eccentric when the valve is in its extreme position. OB″ will give the motion of the valve from its mid position. If the eccentric were in the position OA′, the valve would be in its mid position, and hence if OA is the position of the eccentric at the beginning of the stroke, the angle A′OA must be the angle of advance. This is designated by δ, and it is the angle through which the eccentric must be turned to move the valve from its mid position a distance equal to the lap plus the lead. From the figure it is seen that OB equals OA sin δ, or calling the throw or eccentricity of the eccentric r, the movement x of the valve will be equal to $r \sin \delta$, or $x = r \sin \delta$. Now suppose the eccentric be moved by the crank through the angle ω, then OB″ is equal to $r \sin (\delta + \omega)$. In this expression ω is the amount the crank has been turned from the dead-point. To study the effect of

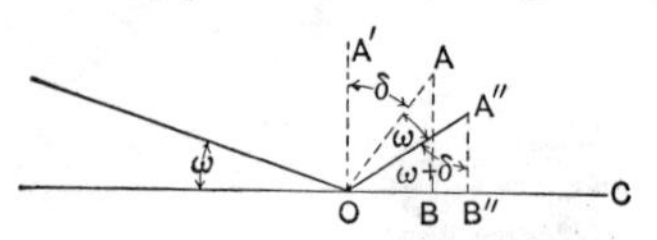

FIG. 183.—Relative Movement of Valve and Crank.

varying the proportions of the different parts, this motion may be plotted in the form of a diagram, which is known as a valve diagram. Different constructions have been proposed by various authors, and those suggested by Bilgram and Zeuner are given hereafter.

Bilgram valve diagram.—On a horizontal line BOC, Fig.

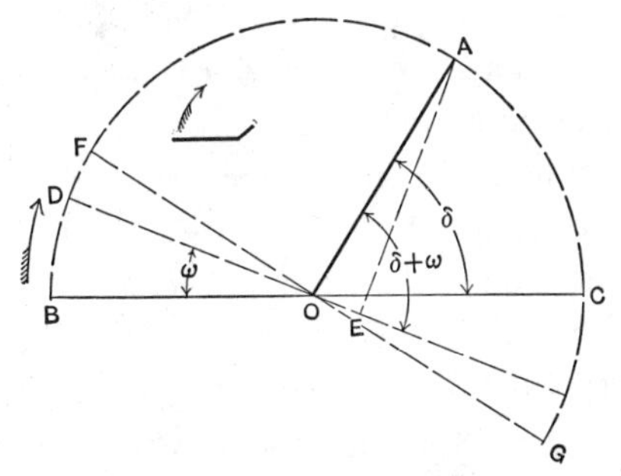

FIG. 184.—Bilgram Diagram.

184, select some point O as center, and from it draw OA at an angle δ with the line OC, and equal in length to the eccentricity r. OB represents the crank at the beginning of the stroke for one end. By "**stroke for a certain end**" is meant that stroke during which admission and expansion occur in that end. The term **return stroke** refers to that stroke during which exhaust for the end just considered occurs. OC is the position of the crank after turning through 180 degrees, and it is from this line that δ is measured in a direction opposite to the assumed rotation of the crank.

Suppose the crank has moved through the angle ω to the position OD, then if from A we drop a perpendicular on OD, or OD produced, this perpendicular AE is equal to the movement of the valve, because AE is equal to AO sin $(\delta + \omega)$, or r sin $(\delta + \omega)$. When ω has such a value that AOE is a right angle, as at OF, AO is equal to the movement of the valve, and as this is the largest possible value of AE, this position OF is the position of the crank for the maximum port opening. As the crank passes this position the length AE decreases, and finally becomes zero when the crank is in the position OA, or this is the position of the crank in which the valve is at its mid position. Passing this point, the valve moves beyond the mid position until the crank reaches OG, when the valve is at the other extreme of its travel and begins to return again, reaching the mid position when the crank is opposite OA. It is seen that the perpendicular let fall from A on any position of the crank gives, at that instant, the motion of the valve from its mid position. This, however, does not give the port opening, for the valve must be moved from its mid position a distance equal to the steam lap before the admission can take place. The port opening is, therefore, equal to the movement of the valve less the lap. If a circle be drawn with A as a center, and with a radius equal to the outside lap, the portion of the perpendicular between the circumference of the circle and the position of the crank, will represent the actual port opening for that position of the crank. Thus, in Fig. 185, AE represents the movement of the valve, but GE the

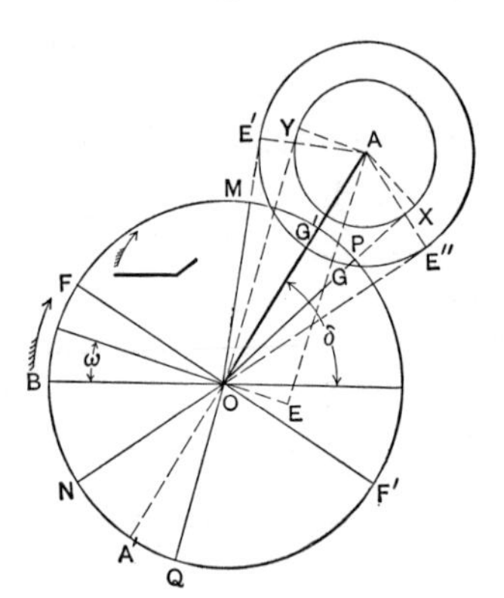

FIG. 185.—Diagram with Lap.

port opening. When the crank reaches the position F, the perpendicular from A is AO, and G'O is the greatest port opening. At M the perpendicular AE' is just equal to the lap, and hence at this point the port is just closed. This, then, is the position of cut off. From this point the valve moves over the port, finally returns to its mid position at A', and then moving on reaches the position N. Here the motion of the valve is AE″, which is just equal to the lap, and if the crank is moved beyond this the valve opens the port. N is therefore the position of the crank for admission. On considering the exhaust side of the valve it is seen that the valve must move a distance equal to the steam lap from the point of cut off before it reaches the mid position, and then, if moved beyond this point a distance equal to the exhaust lap, the exhaust port will open. Hence, if AX represents the exhaust lap, OP is the position of the crank at which release occurs; that is, the point at which the exhaust port opens. This port continues to open until the position OF′ is reached, when the maximum opening occurs, and after this the opening grows smaller until at Q the movement of the valve AY is just equal to the exhaust lap, and hence the exhaust is closed and compression begins.

Negative lap.—If the valve laps over both ports, as shown previously, the laps are both positive. If, when the valve is in its middle position, it is so constructed that it does not cover

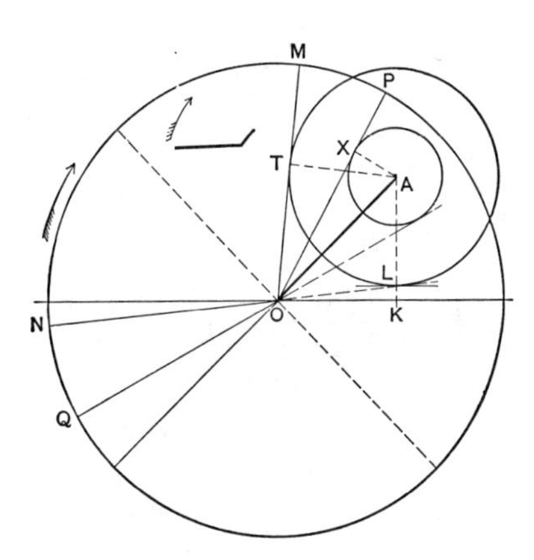

FIG. 186.—Negative Lap.

the port, the amount by which it is uncovered is called negative lap. This sometimes occurs on the exhaust side of a valve, and to give practice in interpreting these diagrams the Fig. 186 is given in which the exhaust lap is negative.

AT is the steam lap, AX the exhaust lap which is negative. When the valve is in its mid position the steam is cut off, but the exhaust is open, hence the cut off occurring at M the release occurs at P before the mid position. The exhaust closes at Q after the mid position. Admission takes place at N. The movement of the valve is AK when on dead center, and hence LK is the lead. The events for the other end of the cylinder take place when the crank occupies positions diametrically opposite those shown for the end just considered, provided, of course, that the laps are the same on both ends.

Although the crank has relatively the same position on each stroke when the different events occur, the piston has different positions on account of the effect of angularity. For this reason the laps are often so changed that the events may take place at the same point of the stroke, although at different crank angles from the centers. The diagram for such an arrangement is shown in Fig. 187, the two large circles being

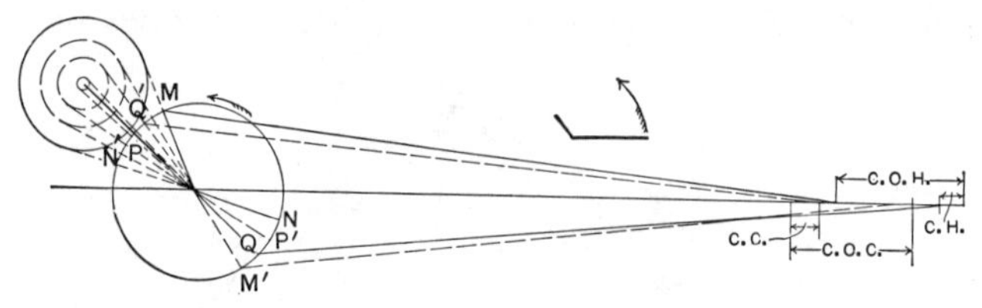

FIG. 187.—Different Laps.

the steam laps and the smaller circles the exhaust laps. The laps are chosen to give equal cut offs and compressions, the release and admission being different on the two ends, the lines belonging to the head end being solid and those to the crank end being broken. M is the cut off on the head end, P the release, Q the compression, and N the admission. For the crank end the primed letters show the same events.

It is to be noted that as the steam lap is increased the cut off becomes earlier and the admission later. With an increase of the exhaust lap, the release is later and the compression earlier. That is, increased lap means increased expansion and compression. It will also be seen that if the angle of advance be increased, without changing any other parts of the eccentric or valve, the events of the stroke will take place earlier. An increase in the eccentricity increases the movement of the valve, and therefore the port opening, as well as changing the events of the stroke, admission and release occurring earlier while cut off and compression occur later.

Problem I.—The valve diagram is useful in the design and the setting of the valves. Suppose, for instance, an eccentric is to be set so that cut off will occur at half-stroke. In this case the eccentricity, lap, and cut off are fixed, and from them we can find the lead and angle of advance. Draw around the point O, Fig. 188, the crank circle to any scale, mark the

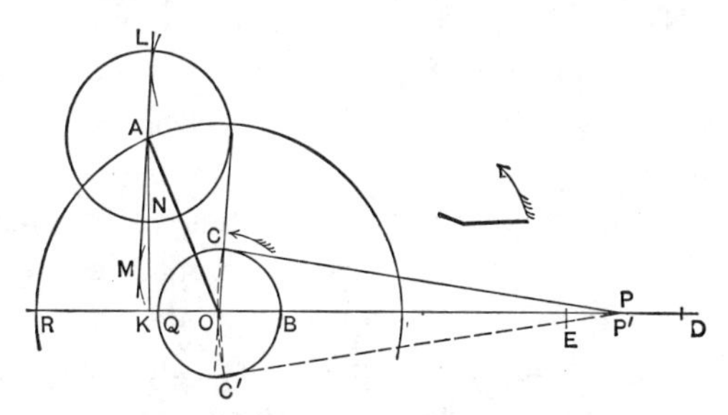

FIG. 188.—Problem I. Setting Eccentric.

points Q and B, and lay off from B the distance BD, equal to the connecting rod to that scale. From D lay off DE equal to QB. This represents the travel of the piston. If we wish to find the crank position for any event of the stroke, the piston position is located between E and D, which in this problem is the middle, and from P as a center describe an arc with a radius equal to BD, the length of the connecting rod, cutting the crank circle at C and C′. If D is the beginning of the stroke and the engine be running over, the crank must be above the center line when cut off takes place; therefore OC represents the position. If cut off occurs at half stroke on the crank end as well, OC′ will be the position of the crank for this event. This diagram shows clearly the effect of angularity.

From the center O, Fig. 188, draw a circle OA to a large

scale, the radius of which is equal to the eccentricity. The center of the lap circle must lie on this circle, and must also touch the line OC, the cut off position. At this point care must be taken to put the lap circle on the proper side of OC. It was noted in describing the Bilgram diagram that the cut off with a positive lap occurred before the mid position. That is, the center of the lap circle is on the opposite side of the cut off line from that of the crank at the beginning of the stroke. Hence in the diagram the lap circle is drawn on the left of OC′, but tangent to it and with its center on the circle OR. Draw a line LM parallel to OC at a distance from it equal to the lap. Where this cuts AR is the center of the lap circle. NK is the lead and AOR is the angle of advance. Cut off on the other end of the cylinder will occur at P′ if the laps are equal, this point being determined from the lower end of CO extended.

Problem II.—Suppose an engine has an eccentric turned solid with the shaft, and it is desired to change the lap of the valve so that cut off will take place at .6 stroke. In this case the angle of advance, the eccentricity, and the cut off are known, and the lap is to be determined. In Fig. 189 OC is

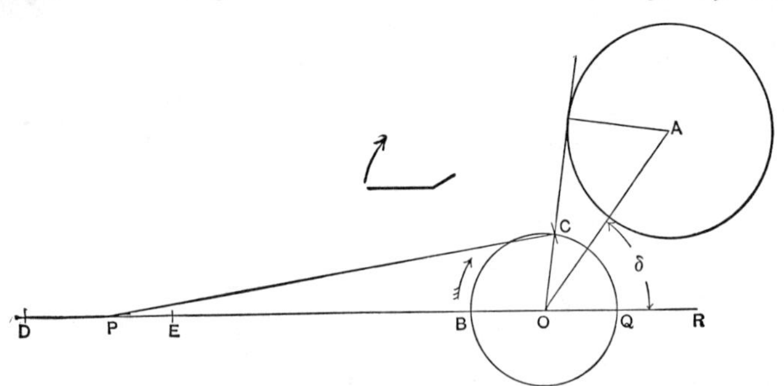

FIG. 189.—Problem II. Changing Lap.

the position of cut off, AOR the angle of advance, and AO the eccentricity. From A draw a circle tangent to OC. The radius of this circle is the required lap.

Problem III.—A new eccentric is to be designed for an engine to give a definite cut off and lead, using the old valve.

What is the eccentricity ? The cut off, lead and lap are known in this case. In Fig. 190 draw OC as the crank position for

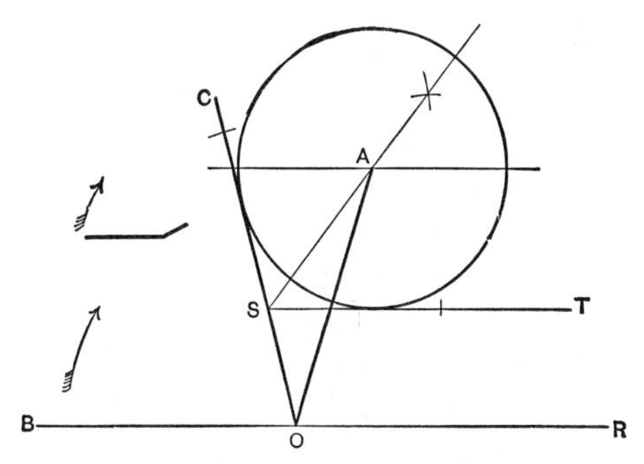

FIG. 190.—Problem III. New Eccentric.

cut off, and the line ST, parallel to OR, at a distance from it equal to the lead. Bisect the angle TSC, and where this bisector is cut by a line parallel to ST, and at a distance from it equal to the lap, is the center A of the lap circle. AO is therefore the eccentricity and the angle AOR is the angle of advance.

Problem IV.—It may be preferred to use the maximum port opening of the engine in the last case in place of the lead, and then the port opening, cut off, and lap would be the known quantities. Draw the position of crank for cut off, Fig. 191, and a circle ZR equal in radius to the port opening. Then find the center of a circle, the radius of which equals the lap, which will just touch OC and the circle ZR. This will give the center A, so that OA is the eccentricity desired.

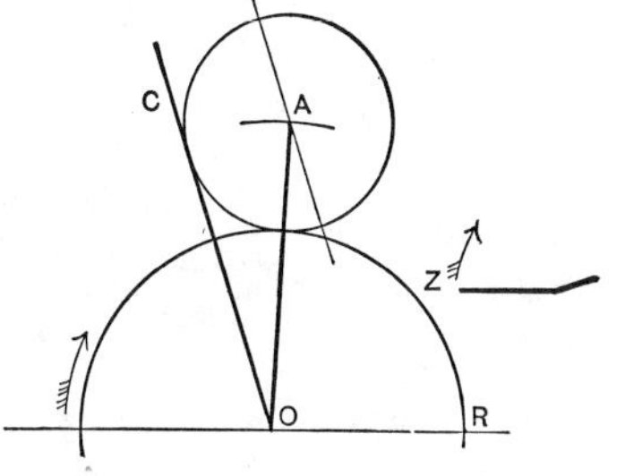

FIG. 191.—Problem IV. New Eccentric.

Problem V.—The general problem in the design of an engine is to find the eccentricity to give a certain port opening, lead, and cut off. The amount of port opening is determined by the size of the engine and the number of revolutions. It is of such an amount that the average speed of the steam is 6000 feet per minute, the steam being assumed to follow full stroke. That is, the area of the steam port is found by the formula

$$a = \frac{2\text{NAL}}{6000}.$$

In this a is the area of the port in square inches; A the area of the piston in square inches; L the stroke in feet; and N the number of revolutions. This holds in cases where the cut off is later than .6 stroke, while for earlier cut offs the denominator is from 18,000 to 24,000. The lead is fixed by practice, varying from 0 to $\frac{1}{2}''$ in locomotives, and to values as high as $1\frac{1}{2}''$ on marine engines. $\frac{1}{8}''$ is a fair value for a stationary engine. Suppose the data for this problem are, the lead $\frac{3}{16}''$, port opening $1\frac{1}{4}''$, overtravel $\frac{1}{4}''$, and cut off .62 stroke. In Fig. 192 lay off the circle OZ with a radius equal to the over-

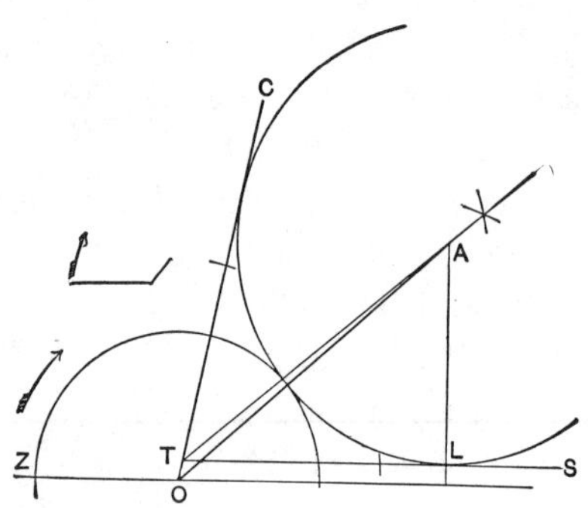

FIG. 192.—Problem V. Valve Design.

travel plus the port opening; that is, $1\frac{1}{2}''$. OC is the position of cut-off and ST is the lead line. Then if we draw a circle which will touch the curve OZ and the two lines OC and ST, its center will give the point A and its radius the lap. AO is the eccentricity. These problems have not considered the exhaust side of the valve, but the same methods will give the data required for that side.

Zeuner diagram.—Another valve diagram, which was suggested by Zeuner and is known by his name, is often used.

If a circle be constructed on a line OF, Fig. 193, as diameter, it can be shown that any line from O to this circle, as OD, will represent the travel of the valve when COD represents the movement of the crank from the dead-point, provided the angle BOF is the angle of advance δ, and OF is the eccentricity r. Since ODF is inscribed in a semicircle it is a right angle, and hence OD equals OF sin OFD.

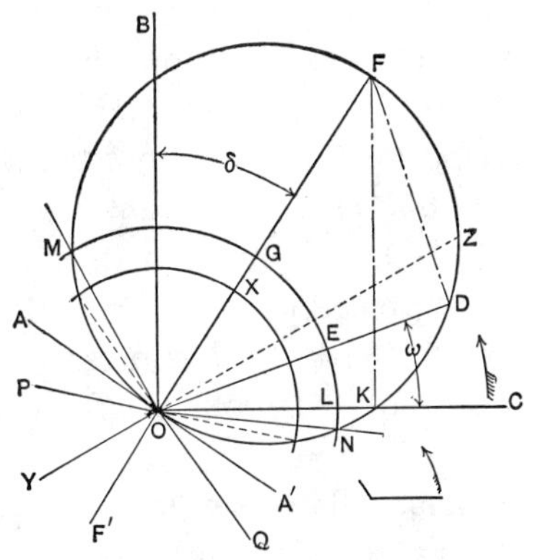

FIG. 193.—Zeuner Diagram.

$$\text{BOF} = \delta,\ \text{DOC} = \omega,\ \text{OF} = r;\ \text{OFD} = 90 - \text{FOD};$$
$$\text{FOD} = 90 - \text{BOF} - \text{DOC} = 90 - \delta - \omega.$$

Hence $\quad \text{OFD} = \delta + \omega \quad$ and $\quad \text{OD} = r \sin (\delta + \omega)$.

This figure is called a **vector diagram** because the length of the **radius vector** OD shows the value of some varying quantity for any direction of the vector. In this case the location of the line shows the position of the crank, and the length of the vector gives the corresponding movement of the valve from its mid position. To construct the Zeuner diagram, two lines OC and OB are drawn at right angles. The first represents the position of the crank for the dead center. From OB lay off BOF equal to the angle of advance, remembering that the angle, when positive, is laid off between OB and OC. On OF, equal to the eccentricity, draw the circle ODFB. Since the port opening is equal to the movement of the valve minus the lap, a circle MGN, of radius equal to the steam lap, must be drawn with O as center. ED is therefore the port opening, and where this reduces to zero either cut off (as at M) or admission (as at N) must occur. The distance LK is the lead and GF is the maximum port opening. When the crank is at a position perpendicular to OF or tangent to the circle, as at A or A′, the valve is in its mid position, and further movement of the crank in the same direction causes the valve to travel to the other side of this position or in a negative direction. This causes the varying vector to trace out the same circle, as the motion is negative, and in such cases the value of the vector is measured in the opposite direction from the pole O. For instance, for the position OY of the crank the movement of the valve is OZ. It can thus be seen that, if OX is the positive exhaust lap, OP is the position of the crank for the opening of the exhaust and OQ is that for compression. This diagram may be used to solve all the problems for which the Bilgram diagram is used.

Reversing gears.—The single eccentric with its rod and valve forming a **valve gear** can be used to drive an engine only in one direction while the eccentric is fixed to the shaft. If, however, we should shift the eccentric from A to A′, Fig. 194, the crank, instead of moving upward from C, would move downward. If two eccentrics were used and we could, by some means, attach either of the eccentric rods to the valve rod, then the same result would be attained.

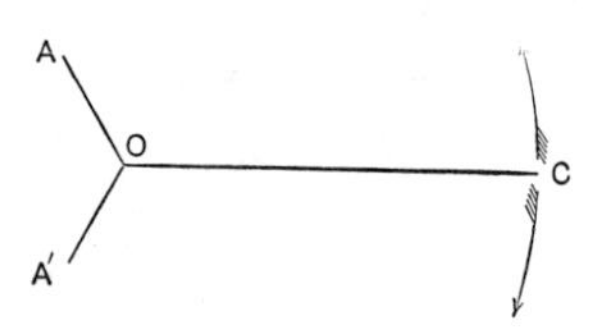

FIG. 194.—Shifting Eccentric.

The Stephenson link motion is a valve gear by which this is accomplished. Two eccentric-rods B,C, Fig. 195, are attached to a curved **link** D by pins E,E. The radius of the link is equal to the length along the eccentric rod from the center of the eccentric to the center line of the link. Within the slot of this link is a **link block** F, over which the link can move. To this block is attached the reverse lever pivoted at G and attached to the valve rod at H. The link is carried by a **hanger** J, shown dotted, attached to a **saddle** K. The hanger is attached to the end of a **bell-crank lever** L on the **reverse shaft** M. If the link is moved to the lowest position, then the eccentric A controls its action, while in its highest position A′ produces the motion and the engine will turn in a direction opposite to that when A controls the valve. Be-

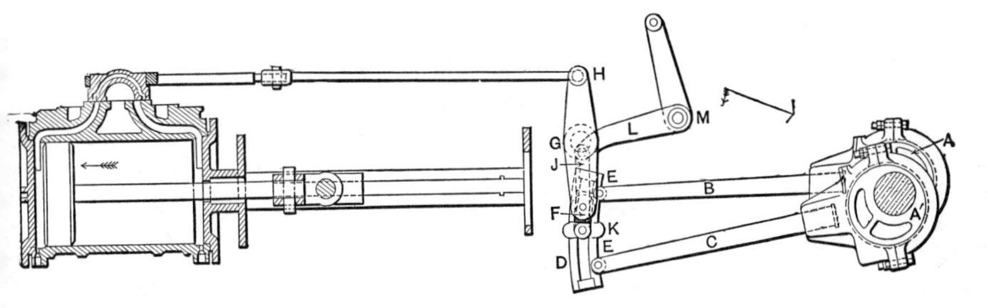

FIG. 195.—Stephenson's Link.

tween these positions, at either of which the link is said to be in **full gear,** the block is acted on by both eccentrics, and on this account the motion is so altered that cut off and the other events occur earlier than would be the case if the link were at the upper or lower position. In no position of the link is the block stationary. In Fig. 196 there are shown two arrange-

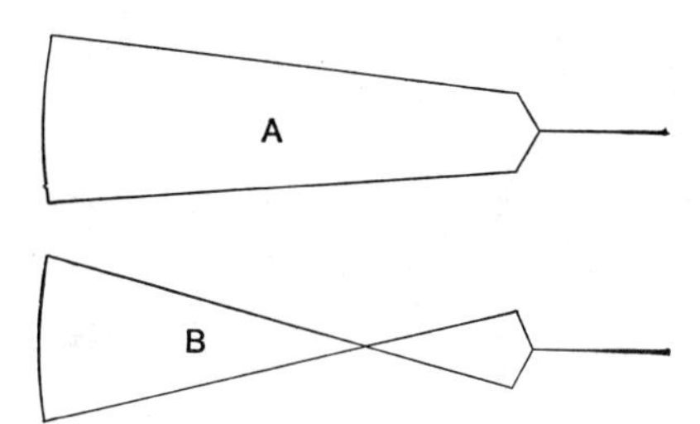

FIG. 196.—Open and Crossed Rods.

ments of the eccentric rods. In the upper one the rods are said to be **open,** and in the lower they are **crossed.** That is, when the crank is on the opposite side of the shaft from the link the rods are said to be open, if, in this position, they do not cross. As the rods, whether opened or crossed, alternate between an open and a crossed position every 180 degrees, it is necessary to choose some position of the crank in defining these conditions, and the position shown above is the one usually adopted.

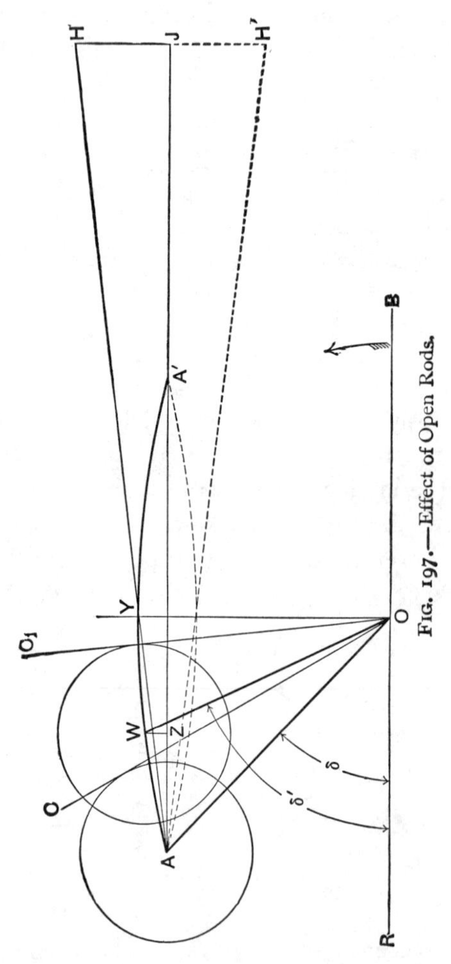

FIG. 197.—Effect of Open Rods.

The effect of open and crossed rods for the Stephenson link motion will be shown from the Bilgram diagram, although the reason for the construction will not be given, the student being referred to a treatise on valve gears for further details and proofs. Lay off the angle AOR, Fig. 197, equal to δ, and OA equal to the eccentricity. From A lay off AJ parallel to OR and equal to the distance between the center of the eccentric and the center line of the link measured along the eccentric rod. For open rods make JH equal to the half-chord of the link to the same scale as AJ and perpendicular to it. Join AH by a straight line and, where this cuts the vertical through O, mark the point Y. Draw an arc of a circle through A and Y, the center being on the line OY, and continue this to A′ on the line AJ. When the link is at its lower position A is the center of lap circle, and the cut off occurs at C, the small circle being the lap circle, and B the position of the crank at the beginning of the stroke. If now the link be raised through one quarter of the length of the chord of the link, the center of the lap circle, for this position of the block, is found by taking the point Z on AA′, such that AZ equals one-fourth of AA′, and at Z erect a perpendicular cutting the arc at W. With this position of the link the motion of the valve due to the two eccentrics is the same as that due to an eccentricity OW and an angle of advance WOR. Hence the cut off occurs earlier than the position OC, that is at OC′ The lead also is seen to be increased. A similar construction is used for any other position of the link.

Since the movement of the valve due to the combined action of the two eccentrics is that produced by an eccentricity OW and an angle of advance WOR, an eccentric having these dimensions may displace the actual gear and is therefore called the **virtual eccentric** for that position of the link. This being the case, any of the problems before used may be applied to the Stephenson motion, for a definite position of the link, if the dimensions of the virtual eccentric be substituted for the actual dimensions. It will be noted that the virtual eccentricity diminishes and the virtual angle of advance increases as the link approaches mid gear, or is **hooked up.**

With crossed rods it is shown in treatises on valve gear that JH′ should be laid off below, giving the dotted curve,

Fig. 197. In this case the cut off is still earlier than at full gear, but the lead has decreased. As the link is moved so that it approaches mid gear the lead increases with open rods, while with crossed rods it decreases.

Gooch link motion.—Another form of reversing valve gear is that shown by the center line diagram in Fig. 198. In this

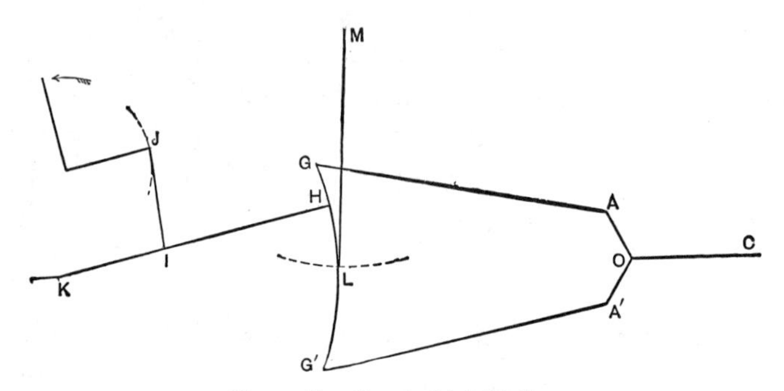

FIG. 198.—Gooch Link Motion.

OC is the crank; OA, OA′, the eccentrics; AG, A′G′, the eccentric rods; GG′, the link; KH, the **radius rod** which is raised and lowered by the hanger JI attached to the bell-crank lever. LM is the **suspension rod** of the link. In this gear the link moves on the rod LM which is pivoted at M, and the block H on the radius rod HK is moved in the link. When K is at the top of the link the eccentric OA controls the motion of the valve rod at H, and, when at the lower end, OA′ controls the motion. As the link has a radius equal to the length of the radius rod HK, the block can be moved up and down without moving the valve when the engine is on either dead center. Hence the lead is the same no matter what the position of the block may be. It is to be noted that, in both this gear and the Stephenson link motion, the eccentricities and angles of advance are shown as having the same values for each eccentric, although this is not always the case.

Radial Gears.—Another method of reversing an engine is by the use of a **radial valve gear.** These are gears in which the motion of the valve is taken from some point in a vibrating link, another point of which moves in a closed curve, while a third point moves in an open curve. They are not used as extensively as the other motions because the parts have to be made very heavy to carry the stresses that are brought on the vibrating links. The gears are simple and do not have many moving parts.

Marshall gear.—Fig. 199 shows a sketch of the Marshall

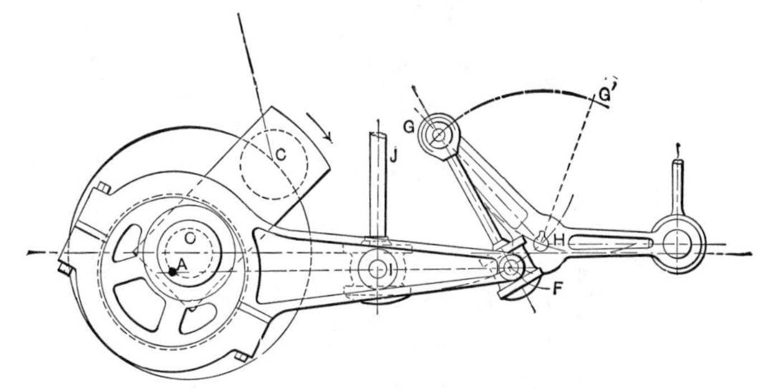

FIG. 199.—Marshall Gear.

gear. OC is the crank; OA is the eccentric directly opposite the crank; AF is the vibrating link; IJ is the valve connecting rod, attached at its upper end to the valve rod; FG is a link equal in length to the arm HG from the upper end of which it is swung. The arm GH turns about the point H.

As the engine moves as shown by the arrow, the point F is caused to move to the left at the same time that it travels downward along the arc. This moves the valve downward, admitting more steam to the upper end of the cylinder and driving the engine. If we shift HG to the right, as shown by the dot-and-dash lines, G would be at G′, and, if the engine moved clockwise, F would travel upward, thus moving the valve in the same direction and closing the port. Therefore the engine could not run in this direction. If it were moving in the opposite direction, the motion of the valve would be downward and hence the engine would continue to run. Thus, by moving the arm HG, we can reverse the engine, and by taking intermediate points we can vary the cut off. This gear gives constant lead.

Joy Gear.—This gear is shown in Fig. 200. It is a compound radial gear, as the vibrating link obtains its motion from another vibrating link and not from an eccentric.

In Fig. 200, OC is the crank and CL the connecting rod.

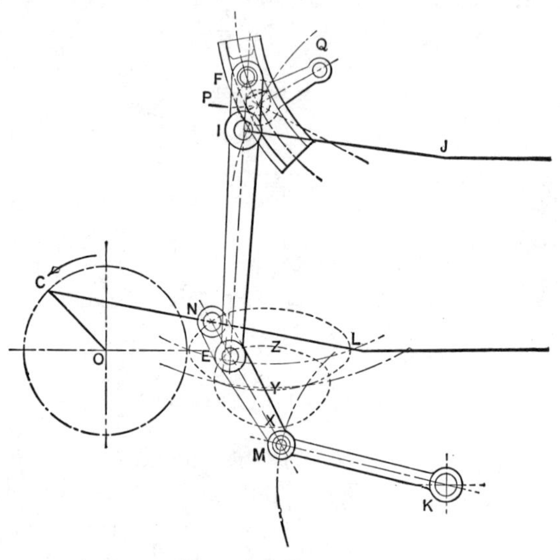

FIG. 200.—Joy Gear.

To a point N on the connecting rod is attached the link NM, which is attached at M to the radius rod MK pivoted at the fixed point K. At the point E is attached the **vibrating link** EI, pinned to the **valve connecting rod** IJ. The point F on EI is attached to a block which moves in a slotted arc. The point I may be between E and F, in which cases the motion of F is opposite to that when I is beyond F. The slotted arc is so mounted that its direction can be shifted from the position shown to that indicated by the dotted arc, or to any intermediate position. In the dotted position the engine will run under, while with the position shown it will run over. In this gear also, the lead is constant.

The reason that the motion of E is not taken directly from N is seen from the curves which show the paths along which N and E move. N moves in an approximate ellipse, while E has a path which extends farther below its position when the crank is on dead center than it does above that position. As the motion of E in the arc of a circle about the center of the slotted arc would give no motion of F, the actual amounts of vertical motion of F above and below the center are respectively YZ and XY, which are nearly the same, giving similar steam distributions to both ends. Y lies on the arc of a circle through P and of radius FE. Had the point N been used, the motions would have been so different that the steam distributions on the two ends would be dissimilar, and the movement of the engine irregular. As the position of E is changed on the link NM, the relative lengths YZ and XY are altered, which allows the steam distribution on the two ends to be proportioned as desired. As the crank turns, the end F is forced to travel in the slotted arc and the side motion of I is transmitted to the valve. The slotted arc is solidly attached to the arm PQ, which is pivoted at P. If the arm is so moved that Q is below a horizontal line through P, the arc will be inclined in the other direction and the engine will be reversed.

Reversing with single eccentric.—It is sometimes convenient to use the central portion of the valve for live steam from the boiler instead of for exhaust steam. The valve is then said to take steam inside, and the effect of this is to reverse the steam distribution. The motion which formerly opened the steam port now tends to close it, and, to give the valve the proper lead, as shown in A and B, Fig. 201, the eccentric must be in one of the positions E or F. To have the engine run in the direction shown at A the eccentric must be placed directly opposite the position in A, as shown at E. The position at F would reverse the direction. It is to be noted that when the engine takes steam inside, the crank leads the eccentric, while with steam outside, the eccentric leads the crank, the smaller angle between the crank and the eccentric being considered.

When the eccentric rod cannot be directly attached to the valve stem a reverse lever is sometimes interposed, as shown by the line diagram at C, Fig. 201. The action of this is exactly

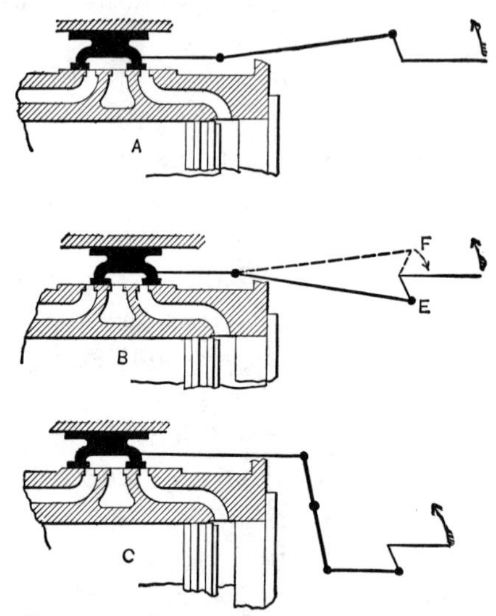

FIG. 201.—Reversing with Fixed Eccentric.

similar to shifting the steam from the outside to the inside of the valve, and hence the crank leads the eccentric. The combination of a reverse lever with a valve taking steam inside would be to bring the relative position of the eccentric and crank to that required with a valve taking steam outside and directly connected to the eccentric.

If a valve be made without lap and lead, the eccentric and crank will be so arranged that the valve will be in its mid position on dead center. If now the steam be taken outside of the valve, the crank will follow the eccentric, while if this be changed so that steam is taken inside, the eccentric will follow the crank, or the engine will be reversed. This method is used in reversing engines which are not run continuously, as is the case with hoisting engines, etc., and when economy is not a primary consideration. In such engines the main valve must be so constructed that it is held on its seat when the steam is brought beneath it, and, moreover, two engines must be used with cranks quartering to prevent the engine from stopping on dead center. Fig. 202 shows the form of reversing valve D,

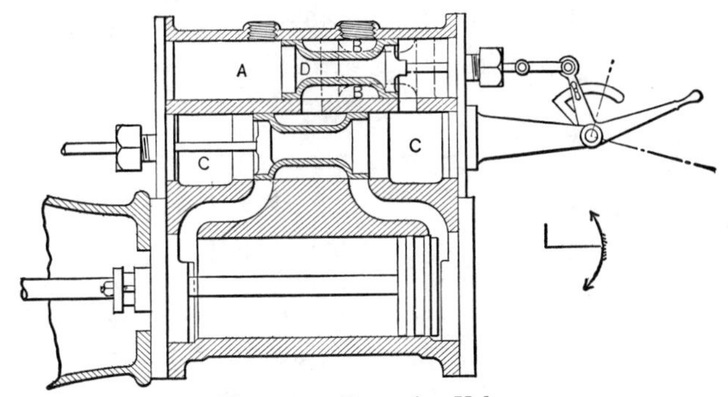

FIG. 202.—Reversing Valve.

used to shift the steam from the outside to the inside. In the position shown, steam from A can enter the space C, while the exhaust space B is connected to the inside space of the main valve. When the reversing valve is shifted to the right, as shown by the dotted lines, steam is admitted to the inside of the valve, while the exhaust space B is connected to the space C. This causes the engine to reverse, as is shown in the small figures.

Valve for single cylinder direct-acting pump.—The valve for a single cylinder pump is of special construction, as the motion of the pump must be used to drive this valve. Fig. 79 partially shows this, steam in M entering beneath the large valve H, shown in section, is driving the piston C to the right. As this occurs, the arm N strikes the tappet O on the rod R, moving the rod. This action moves an auxiliary valve which admits steam in back of the piston valve S, and forces it to the right moving the main valve H, and so permitting steam to enter behind the right-hand end of the piston C, forcing it back. This main valve H and the auxiliary valve T are shown in Fig. 203. The valve T is attached to the rod R, and contains within the hollow rectangle formed by its sides, the valve H. The valve T is in reality two valves, united at the ends by UU, with exhaust cavities in the lower faces as shown. These valves move over the spaces *a*, *b*, *c*, *d*, *e*, Fig. 203, which shows the seats of these valves, *ff* being the steam ports for the main valves H, and *g* the exhaust port. The spaces *h*, *h*

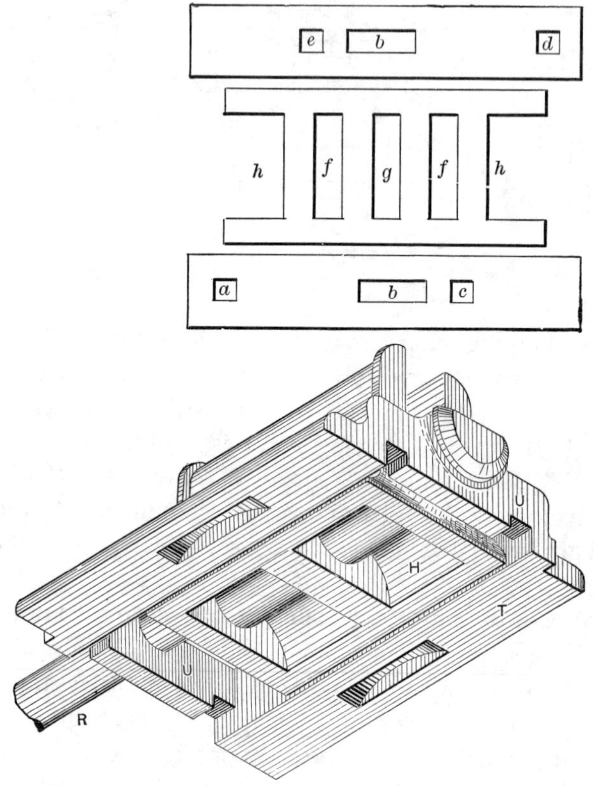

FIG. 203.—Pump Valve.

are shown at M in Fig. 79. The spaces *b*, *b*, are connected to the exhaust space of the pump; *a* and *e* connect to the left-hand end of the space in which the auxiliary piston S moves, while *c* and *d* connect to the right-hand end. If now the rod R moves T to the right, *a* is uncovered and opened to the steam, while *d* and *e* are covered, and *c* and *b* are connected by the front exhaust cavity. This, then, forces S, Fig. 79, to the right, and, as explained, reverses the pump. When the left-hand tappet is struck the valve T is moved to the left, and then *a* and c are covered, *e* connected to the exhaust *b*, and *d* is open to the steam. This forces the auxiliary piston to the left. The main valve H fits between the two parts of S, as shown in Fig. 79, and is driven by them. To render the construction simpler the steam-passages *a*, *c*, *d*, and *e* do not extend to the end of the steam chest, but enter near the middle of the auxiliary piston, the steam entering the end through the groves and holes in the piston, which are shown by the dotted line in Fig. 79.

CHAPTER VII.

INDICATING AND GOVERNING.

The Steam-engine Indicator.—This apparatus is intended to show graphically the variation of pressure in the cylinder of an engine. It consists of a piston A, Fig. 204, working within a cylinder B, and to which is attached a spring which is fastened to the cap of the cylinder. As the compres-

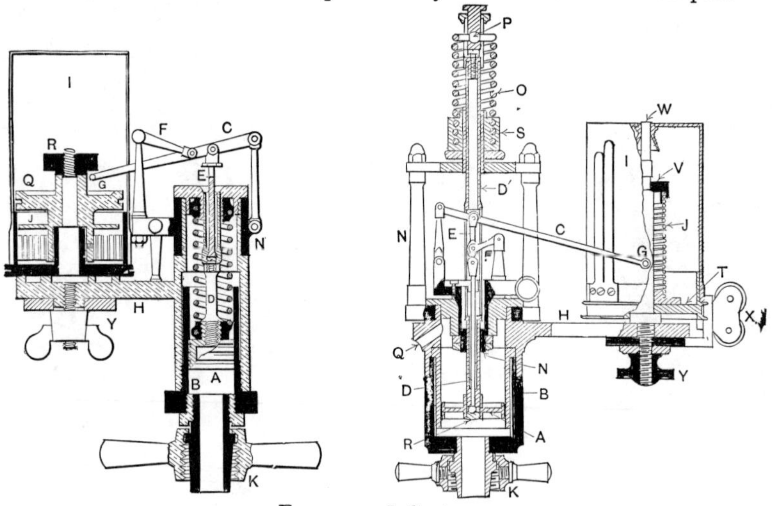

FIG. 204.—Indicators.

sion of a spring is proportional to the amount of pressure exerted, the movement of the piston of the indicator will be the measurement of this pressure. The piston is attached to the **pencil bar** C by the piston rod D and the link E, the arrangement of the points of attachment on the bar and the length of the links E and F, as well as the length of C, being such that the pencil at G moves in a straight line. To the cylinder is attached an arm H, which carries a **drum** I. This drum is

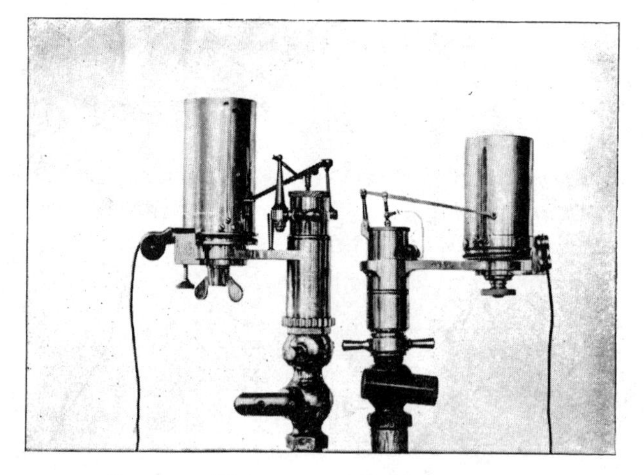

Fig. 205.—Thompson and Crosby Indicators.

(*To face page* 191.)

mounted on an axis, and is provided with a spring J, which draws it back to the original position after it has been rotated.

The indicator is attached, by a coupling K, to a cock on the cylinder of the engine, and, on opening this cock, the steam will act on the piston and cause it to move as the pressure in the cylinder rises and falls. The drum I is attached by cords and levers to the cross head of the engine, and, as the piston of the engine moves, this drum is turned in a similar manner, being moved through a fraction of its total motion when the piston of the engine moves through the same fraction of its stroke. The upper portion of the indicator containing the **straight line motion,** as the links or bars are called, is mounted on the piece N, so that the pencil can be pressed against a piece of paper on the drum I. In this way the movement of the piston is recorded. In the Crosby new indicator, shown on the right, the piston spring O is fastened to the piston by the rod D′, which screws into a threaded projection on A, holding a ball between a spherical cavity in the end of the rod and one on a small screw R on the piston, thus holding the piston in place. The upper end of the rod D′ is connected to the spring O through the ball P on the spring. The lower end of the spring O is attached to the nut which screws on the part N. An opening Q above the piston is provided to insure the pressure being that of the atmosphere at that point.

In the Thompson indicator, on the left, both ends of the piston spring are screwed as described above for the upper end of the Crosby spring. The Crosby drum spring J is fastened to the drum base T at its lower end, and to a nut V above. By removing the drum and raising this nut from the square portion of the axis W it may be turned, thus making the tension either greater or less. The drum spring in the Thompson indicator is hooked over projections at both ends and is tightened by the knurled cap Q and jam nut R. The pulleys X guide the drum cord to its point of attachment. The thumb nut Y clamps the pulley arm in any desired position. Fig. 205 shows an exterior view of a Thompson and of an inside spring Crosby indicator mounted on their attachment cocks. This cock, a section of which is shown in Fig. 206, is so made that, in the position shown, steam is admitted to the cylinder of the indicator, while turning through a right angle cuts off the steam and opens this cylinder to the atmosphere. In this position, air being on each side of the piston, the spring is under neither compression nor tension, and the line drawn by the pencil is called the **atmospheric line.**

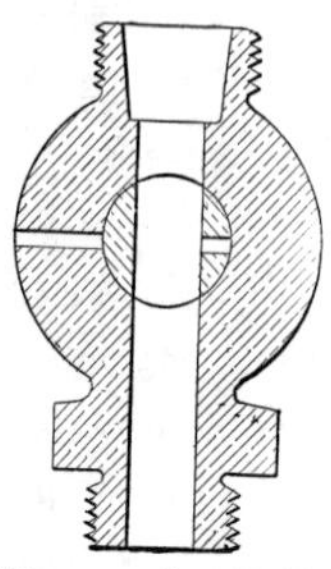

FIG. 206.—Indicator Cock.

Separate indicators should be connected to each end of the cylinder, but often the two ends are connected by pipes to a cock which will connect an indicator to either end of the cylinder or to the atmosphere, being in reality a four-way cock. This practice is not to be followed, except when necessary.

Indicator Rigs.—As the motion of the indicator drum is only about four inches, it is necessary to drive the drum from some form of indicator rig, which is an arrangement of levers

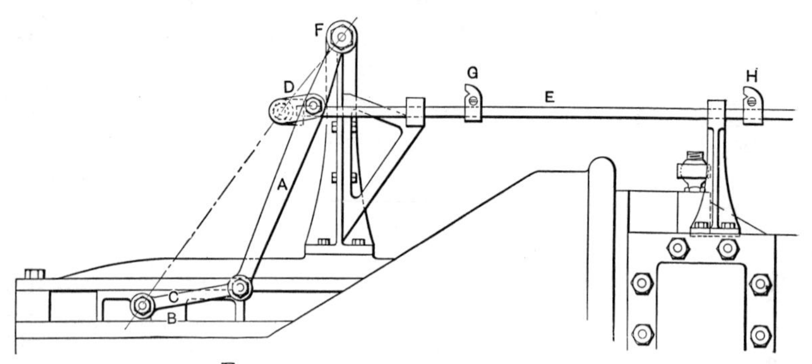

FIG. 207.—Pantograph Indicator Rig.

or wheels by which the motion of the cross head is reproduced to some smaller scale. By reproduction is meant that, when the cross head has moved any fraction of the stroke from dead center, the drum has been moved the same fraction of its total motion from one of its extreme positions.

There are several methods of accomplishing this. The reducing rig shown in Fig. 207 is one which gives an exact reproduction, giving a card 4″ long from a stroke of 16″. The vibrating lever A is driven by the cross head B through the link C. The link D drives the rod E. Since the lengths of the links D and C have the same ratios as the segments of the arm A, or as the distances from the center F to the line of action of the rod E, and from the center F to the line of the cross head, the rod E has a motion similar to that of the cross head. The rod E is extended to the two indicators which are attached at G and H, thus doing away with the necessity of long strings which cause errors on account of stretching.

Fig. 208 is another form of reducing rig. A vertical pin F, attached to the cross head, causes the block G to move with the cross head, while permitting free vertical motion. G is pivoted to OB at B. In this form the motion is slightly in error, due to the vertical movement of the point of attachment of the string. This, however, is slight and is negligible when we consider that there is some stretching of the string. In this case the horizontal motion of the point A is always $\frac{OA}{OB}$ times the motion of B, and hence, not considering the effect of the vertical movement, this is correct.

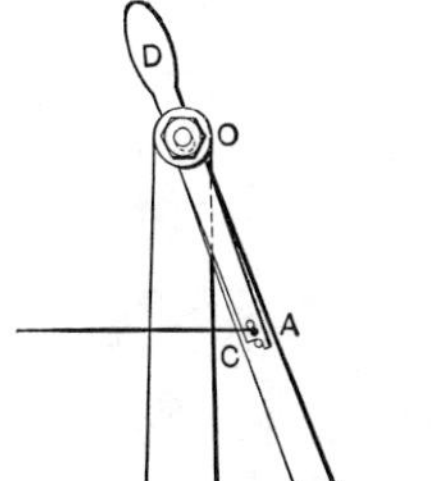

FIG. 208. — Approximate Reducing Motion.

The arm OA is free to move on OB, being held against the pin C by the tension of the indicator string. When it is desired to stop the indicator the handle D is gripped and pulled to the left and then lifted. The slot at O permits this, and the indicator spring will pull the arm to the left, keeping it out of the way until again required. To throw the rig in gear the handle is again lifted, forced to the left, pushed down, and then allowed to move back until the arm is struck by the pin C, when the handle is allowed to go free. In all rigs there should be some method to connect or disconnect the rig or indicator.

Other rigs are manufactured using gear wheels or drums of

different diameters, or using a modification of the pantograph, in all of which, however, the aim is to get an exact reproduction.

Indicator diagrams.—The action of the indicator is as follows: Suppose the indicator drum be connected through the reducing rig to the cross head, which is at the beginning of the stroke, and steam be admitted beneath the indicator piston. This compresses the spring an amount corresponding to the pressure, moving the pencil to the point A, Fig. 209.

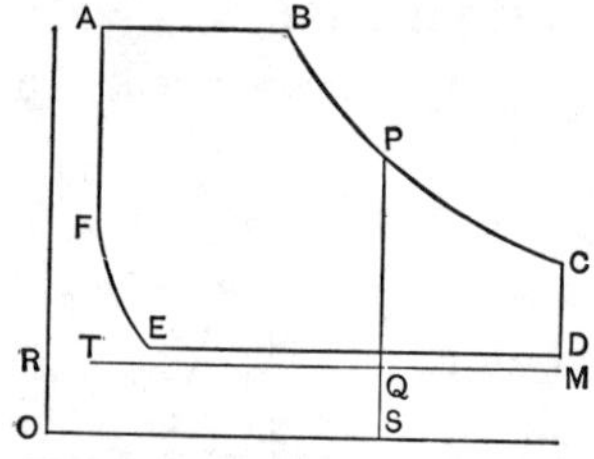

FIG. 209.—Indicator Diagram.

As the engine piston is moved by the steam, the drum is moved by the cross head, so that the horizontal line AB is drawn to the point B where the steam is cut off by the valve. After the point of cut off the steam expands, the pressure falling until the end of the stroke is reached when the exhaust occurs at C. At this point of release, the valve opens the engine cylinder to the atmosphere and the steam is exhausted, the pressure falling to D. As the piston returns the steam is driven out, and the line is drawn to the point E, where the exhaust is closed. As the piston is further moved, after the point of compression, the steam behind the piston in the clearance space is compressed, the pressure rising until F is reached, when the valve again admits steam to the cylinder, raising the pressure to A. F is called the point of admission. The various lines of the diagram also have specific names. FA is called the **admission line**; AB is the **steam line**; BC is the **expansion line**; CD is the **exhaust line**; DE is the **back pressure line,** or, in certain cases, the **vacuum line**; EF is the **compression line.** If the indicator cock is now turned so that the atmosphere acts below the piston of the indicator, we can draw the line TM. This is called the **atmospheric line,** and should be drawn after taking the indicator card, as the length of the spring varies with the temperature. The atmospheric line TM is of the same length as the card, and represents to some scale the stroke of the piston, or the volume swept through by the piston. For instance, the distance TQ, which is two-thirds of TM, represents the condition when the piston has traveled two-thirds of its stroke, the volume swept through being represented by this same distance. To represent the total volume in the cylinder at any point of the stroke, the **clearance volume** must be added to the volume swept through by the piston. To represent this, on the figure, the distance TR is laid off equal to TM multiplied by the percentage clearance. This then gives us the zero of volume. Since the atmospheric pressure is about 14.7 pounds per square inch above the absolute zero of pressure, the distance RO must be laid off equal to that pressure to the **scale of pressure.** By the scale of pressure we mean the number of pounds per square inch which will compress the spring sufficiently to cause the pencil to be moved one inch. Springs are provided, so that this amount may be 10, 20, 40 pounds, or any amount desired within certain ranges. **A forty spring,** for instance, will allow a pencil motion of two and a half inches above the atmospheric line when acted on by steam at 100 pounds pressure. This scale of pressure is also called the **spring scale.**

The absolute pressure and the volume at any point in the stroke are now known, the pressure PS and the volume OS, for instance, being the condition for a movement of the piston TQ.

The expansion curves from many engines have been examined and have been found to agree quite closely with a curve called the rectangular hyperbola. This curve has the property that the product of the distances of any point from the line of zero pressure and from the line of zero volume is constant. This may be written as

$$SP \times OS = \text{constant, or } pv = K.$$

When the expansion line departs from this line, there is some adjustment necessary about the engine. It is to be remembered that there is no theoretic basis for this agreement, and the employment of the rectangular hyperbola rests solely upon the facts that it is a curve easily constructed, and that it is one which agrees closely with the expansion curves from the best engines.

The actual card from an engine differs from that shown in Fig. 209. If the steam passages and the openings of the valves are not large enough, the admission line is inclined, and, when the simple D slide valve is used, its closing is so gradual that, in place of the sharp point of cut off at B, we have a rounded curve. The point of release usually occurs before the end of the stroke, and is so gradual that CD is inclined and not vertical. The exhaust ports and passages being small, the line DE is raised, and at E there is a rounded corner. These effects are shown in Fig. 210, the rectangular hyperbola being shown dotted in.

To draw the rectangular hyperbola through any point, as X, Fig. 211, draw from X a vertical line XH, and from the intersection of the lines of zero volume and pressure draw any inclined line JO. From the point where this cuts a horizontal through X, draw the vertical JK. From the point of intersection of OJ and XH draw the horizontal SK. The point K, where this cuts JK, is another point on the curve. Proceeding in this manner we can find other points, as Y and Z.

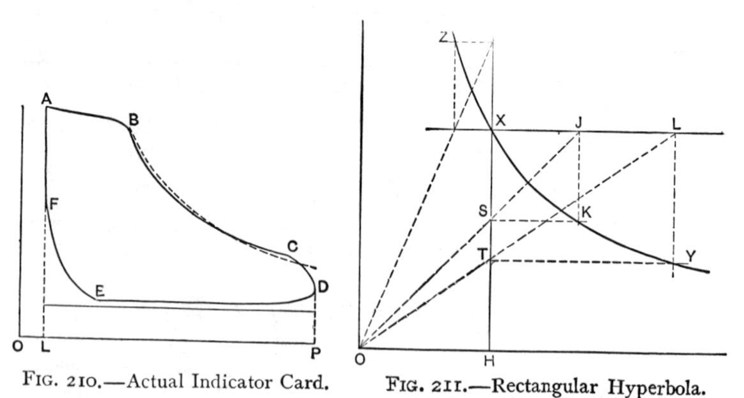

FIG. 210.—Actual Indicator Card.

FIG. 211.—Rectangular Hyperbola.

By the indicator-card we can tell whether or not the valves are set correctly, as the card shows the actual distribution of the steam. Fig. 212 shows a late steam admission and release, as well as late compression. If the eccentric is shifted ahead on the shaft the events will be made earlier and these errors

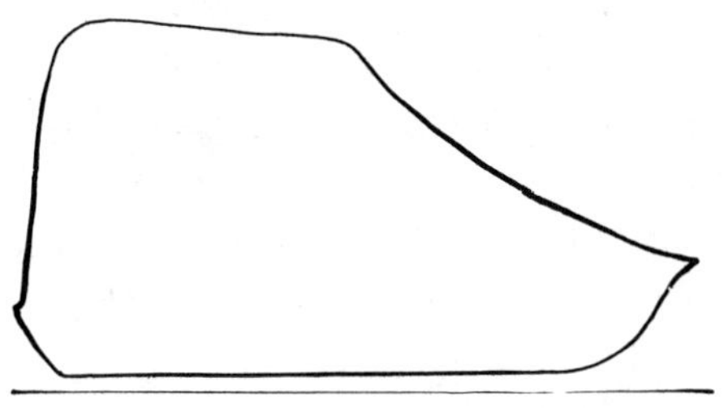

FIG. 212.—Late Admission and Release.

will be corrected. The effect of small exhaust port is shown in Fig. 213 by the gradual drop from the point of release and

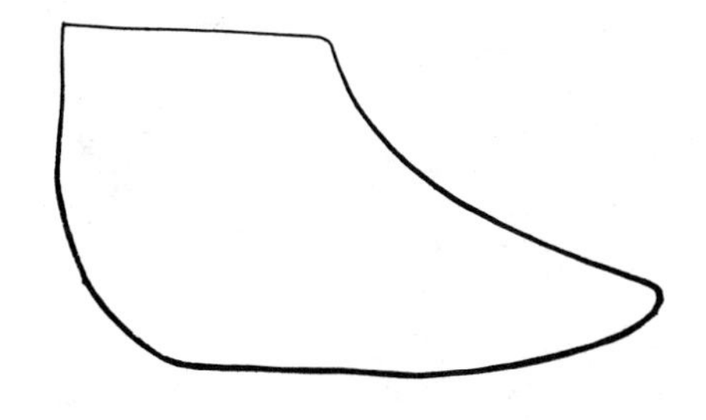

FIG. 213.—Small Exhaust Port.

the height of the exhaust line above the atmospheric line. This latter may also be due to a small exhaust pipe.

When the steam enters the cylinder there is condensation of part of the steam, which is called initial condensation, as explained in Chapter V. As the steam pressure falls during expansion, the heat contained in the cylinder walls and in the condensed steam causes a re-evaporation, which, as the amount of steam in the cylinder at these later points is increased, produces a gradual rise in the pressure. This phenomenon is shown by the expansion curve rising above the rectangular hyperbola, as shown in Fig. 210. This same result can be produced by a leaky valve, allowing steam to enter during expansion, which is more frequently the cause of this rise. The expansion line falling below the hyperbola generally shows a leaky piston, as the steam can pass directly to the exhaust side of the piston. Continued condensation would produce this, but this condition rarely occurs.

Valve setting by Indicator cards.—If the indicator be attached by a four way cock to either the atmosphere, the crank end of the cylinder, or the head end, two diagrams can be drawn on the same card, as shown in Fig. 214. On the

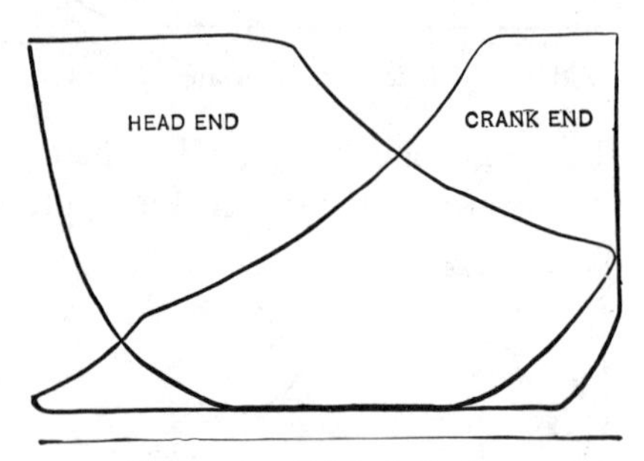

FIG. 214.—Original Cards.

cards here shown the cut off and release are late on the head end, while those on the crank end are early, the compression being too early on the head end. This shows that the valve has been moved on the stem so that the laps are not correctly proportioned. If we move the valve on the stem, or lengthen the stem, making the steam lap for the head greater, and that for the crank end less, the cut offs will be equalized on the two ends, shortening that on the head and making that on the crank later. The exhaust laps are changed differently; that on the head end being shortened and that on the crank end lengthened. This will change the release and compression, so that we get the cards shown in Fig. 215.

The shifting of the valve on the stem is generally not sufficient to bring the cards into the shape shown in this figure. The indicator cards should be made as nearly identical on the two ends as possible, the valve being then at its right position on the stem. The eccentric should now be shifted on the shaft to bring the admission line practically vertical.

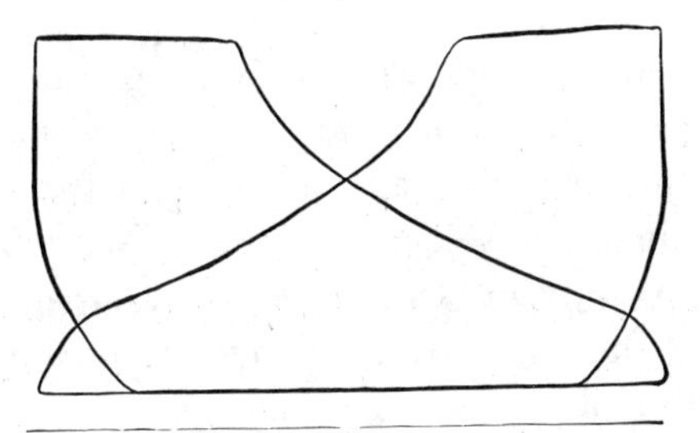

FIG. 215.—After Adjusting Engine.

This applies to any engine in which the valves are driven from a single eccentric.

Fig. 216 shows a card from an engine in which, on the

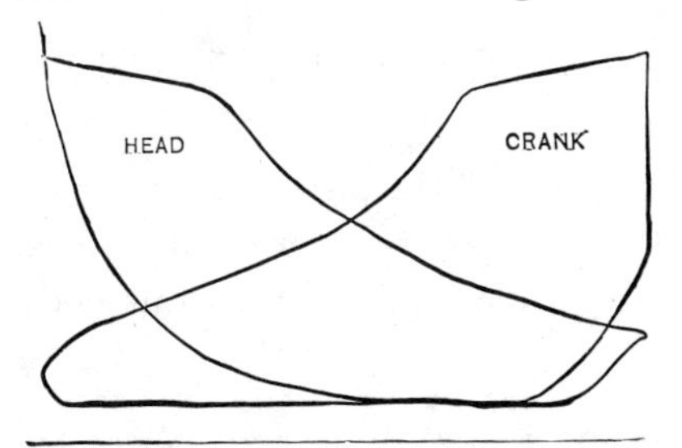

FIG. 216.—Incorrect Exhaust Lap, Head End.

head end, the compression is too great and the release too late. The exhaust lap for that end is reduced to make the compres-

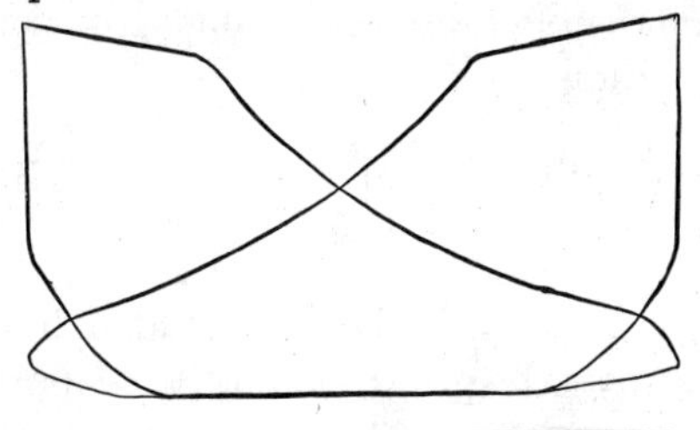

FIG. 217.—After Correcting Exhaust Lap.

sion smaller and the release earlier, giving the cards shown in Fig. 217. The drop in steam line is due to the small motion of the valve, and to correct this a larger eccentric is made, together with a new valve, on which the laps are increased in the same proportion as the eccentric has been. This gives a card as shown in Fig. 215, correcting the wire drawing shown at the admission and release. When pressure is reduced by this throttling action, as the effect of the narrow opening is called, the steam is said to be **wire drawn.**

Measurement of Power by Indicator Cards.—As the heights of the indicator card measure pressures per square inch when the piston occupies different positions, the mean height of the curve ABD, Fig. 210, from OP represents the mean pressure pushing the piston forward, and the mean height of the curve DEA represents the pressure against which the piston is moved backward. Let s be the spring scale and let $\frac{\text{ABDPLA}}{\text{LP}} \times s = p$, which is the mean forward pressure per square inch in that end of the cylinder.

If a equals the area of the piston in square inches, pa equals the mean total pressure during the outward stroke. If the stroke is l feet, the gross work done in this end of the cylinder is pla.

Let $\frac{\text{DEFLPD}}{\text{LP}} \times s = p_0$ and then p_0la will equal the work done by the piston in forcing the steam out of this end of the cylinder on the return stroke, and hence $(p - p_0)\,sa$ will be the net work done, in foot pounds, on one side of the piston per revolution of the engine.

$$p - p_0 = \left(\frac{\text{ABDPLA}}{\text{LP}} - \frac{\text{DEFLPD}}{\text{LP}}\right) s = \frac{\text{ABDEFA}}{\text{LP}} \times s;$$

or, the area of the card ABCDEFA divided by the length LP when multiplied by the spring scale represents the **mean effective pressure,** which is usually written **M.E.P.** Calling the M.E.P. p, we have pla as the work done in this end of the cylinder per revolution. The work done per minute by one end is then $plan$ measured in foot pounds.

To compare engines Watt assumed that a horse could do 33,000 foot-pounds of work per minute, and when an engine does this amount of work per minute, it is said to develop **one horse power.** Thus, the horse power developed by one end of an engine is $\frac{plan}{33000}$.

In the crank end of the cylinder the area of the piston is reduced by the area of the piston rod, and a for the crank end is, therefore, smaller than that for the head. Since the mean effective pressure is usually different for the two ends of the cylinder, the horse power of an engine is given by the expression

$$\text{Horse power} = \frac{(p_c a_c + p_h a_h)ln}{33000},$$

$\frac{la}{33000}$ being constant, for a given engine, is called the **engine constant** for the end having an area a.

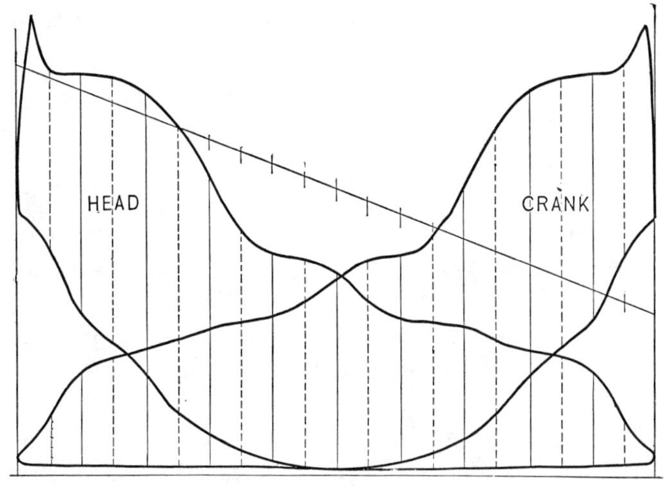

FIG. 218.—Measuring Horse Power.

To determine the horse power from an indicator card, the cards shown in Fig. 218, taken from a 6″ × 8″ engine running at 300 revolutions, will be worked out. The cards are quite wavy, which is usually the case in cards from high speed engines when the pencil levers are heavy and a light spring is used. This is due to vibrations of the indicator spring, and although the lines drawn do not agree with the true expansion line of the steam, the piston so oscillates above and below the true position that the area under the wavy curve is nearly equal to the true area. The areas of the cards are best determined by a planimeter, which is an instrument for measuring areas. Where one of these is not accessible the mean height may be found as follows: Draw two lines at right angles to the atmospheric line at the extremities of the card, divide the distance between into ten equal parts, and, at the points of division, draw lines perpendicular to the atmospheric line. This divides the card into ten trapezoids, and the middle line of each of these is drawn as shown by the dotted lines. The area of each of these trapezoids is proportional to the middle line, since the distances between their parallel sides are equal. The mean height of the figure is therefore the mean of these middle lines, which is found by laying them off consecutively on a straight line and dividing the total length by ten. It is not necessary to draw the solid lines, as the result can be accomplished by dividing the base into twenty parts and drawing perpendiculars at the odd numbered points from either extremity; that is, the first, third, fifth, etc.

In the cards of Fig. 218 the head card gives 9.2 inches for the sum of the lengths of the middle lines and the crank card 9.4 inches. The mean height is then .92 inch for the head card and .94 inch for the crank card. The spring scale is 40, so that the mean effective pressures are 36.8 pounds per square inch for the head and 37.6 for the crank.

The engine turns 300 times per minute and has a piston rod of $1\frac{1}{2}$ inches diameter. From these the following is determined:

$$a_h = \frac{\pi}{4} \times 6 \times 6 = 28.27,$$

$$a_c = 28.27 - \frac{\pi}{4} \times \frac{3}{2} \times \frac{3}{2}$$

$$= 28.27 - 1.77 = 26.5,$$

$$l = \tfrac{8}{12}$$

$$n = 300.$$

$$\text{Engine constant (head end)} = \frac{28.27 \times \frac{8}{12}}{33000} = .000571.$$

$$\text{Engine constant (crank end)} = \frac{26.50 \times \frac{8}{12}}{33000} = .000535.$$

I.H.P. (head end) $= .000571 \times 36.8 \times 300 = 6.29.$
I.H.P. (crank end)$= .000535 \times 37.6 \times 300 = 6.03.$
Total I.H.P. $= 12.32.$

This horse power is called the **indicated horse power** of the engine because it is determined by the indicator. Friction of the various parts will consume power, so that the amount delivered from the shaft is smaller than the indicated power. The amount lost in the engine is termed the **friction horse power** and the remainder, which is delivered from the engine, is called the **delivered horse power,** and because it is determined in some cases by the Prony Brake, it is also termed the **brake horse power.** These several horse powers are represented by the symbols **I.H.P., F.H.P., D.H.P.,** and **B.H.P.**

A Prony brake consists of a band A, Fig. 219, encircling the fly wheel of an engine, having attached to it a weighing arm B which prevents the band from turning, and also transmits the force produced by the friction between the band and the wheel to the scales C. To prevent the band from being destroyed, wooden blocks D are fastened to it between it and the wheel. The bolt E is used to change the tension of the band and alter the friction. Water is introduced into the rim of the wheel by the pipe F and, after abstracting some heat from the rim, flows out at G.

The force at the end of the arm B acting at a distance of R feet from the center of the wheel, the radius of the rim being

r feet, is $\frac{r}{R}$ times the force of the friction which acts over the surface of the rim. The force on the end of the arm being F, the sum of the friction forces at the surface of the rim is $\frac{FR}{r}$.

As the force F is continually acting, it may be clearer to consider it to be produced by a load F on a cable which is being wrapped in a groove of radius R on the wheel.

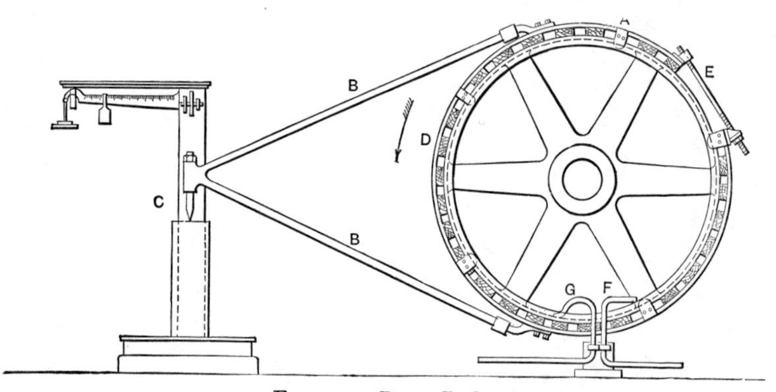

FIG. 219.—Prony Brake.

The distance through which this force F acts for one minute is $2\pi Rn$, where n, as before, is the number of revolutions; hence the work done is $2\pi FRn$. The brake horse power is therefore $\frac{2\pi FRn}{33000}$, $\frac{2\pi R}{33000}$ being the **brake constant.**

Fig. 220 shows a simple form of brake made by a rope wrapped around the wheel, the two ends A being fastened to a frame carried on a platform scale. The middle B of the rope is attached to a tightening device carried by the same frame and the difference in tension is weighed by the scales. The arm on which this force acts is the distance from the center of the shaft to the center of the rope.

The difference between the indicated horse power and the brake horse power is the friction horse power.

$$\text{I.H.P.} - \text{B.H.P.} = \text{F.H.P.}$$

The **mechanical efficiency** of an engine is the ratio of the brake horse power to the indicated horse power.

$$E = \frac{B.H.P.}{I.H.P.}.$$

Governing.—As the load against which an engine acts is continually changing, there must be some way of varying the power which is developed in the cylinder. If this were not

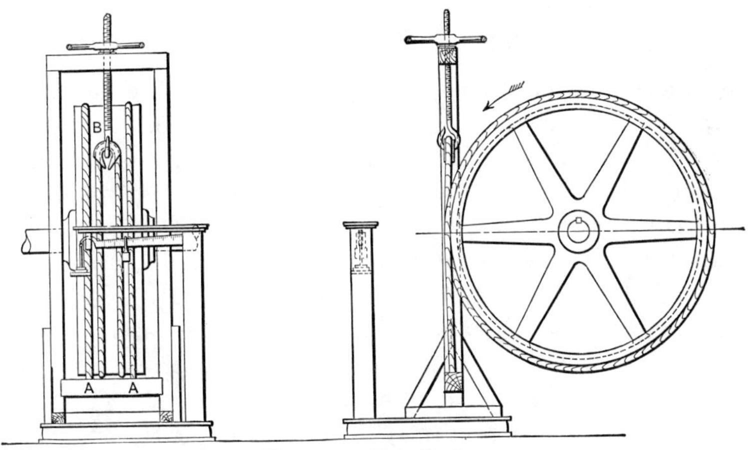

FIG. 220.—Rope Brake.

done, the excess of energy at a time of low load would be used in speeding up the fly wheel, which would finally break, owing to the great centrifugal force at high speeds. As the speed increases, due to the decrease of load, an apparatus called the **governor** acts on the steam supply or valve gear, so that the area of the indicator card is made smaller, thus reducing the indicated power. If the demand for power increases the governor again acts so that the card is larger, more power being developed. An excess or deficiency of load would cause the engine to change its speed, and it is this change of speed, small though it may be, which causes the governor to act.

There are two general methods of governing; that by **throttling,** and that by a **change in cut off.** The throttling method consists of closing a valve in the steam line so that the pressure of the steam entering the cylinder is reduced. The indicator cards shown in Fig. 221, exhibit the effect of

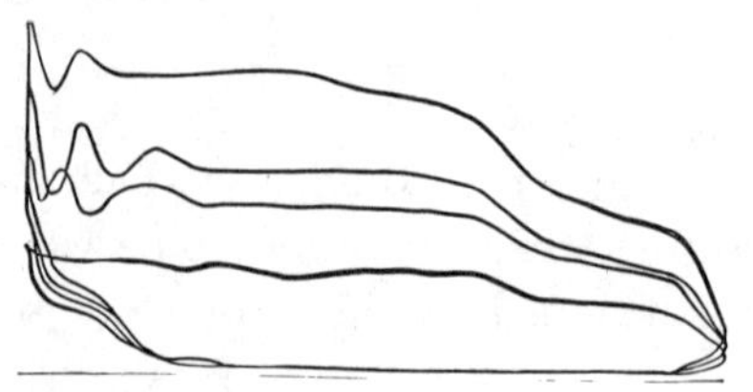

FIG. 221.—Effect of Throttling Governor.

gradually reducing the load on an engine with a throttling governor, the largest card being taken at full load. It is to be noted that the events of the stroke occur at the same piston positions for each card, but as the steam is throttled, the steam line is lower. Although the volume at cut off is the same on each card, the weight of steam used is decreased, because as the steam is throttled through the governor valve it expands so that it occupies a greater volume. Throttling also tends to raise the temperature of the steam above that corresponding to the pressure, which is called **superheating.**

The method of governing by changing the cut off, which gives the name of **automatic cut off engine** to those on which it is found, is accomplished by the governor acting on the valve gear, so that the events of the stroke occur at different points in the stroke. Fig. 222 shows a set of cards taken as the load on an engine is decreased. It will be noticed that, as the load is taken off, the cut off, release, compression, and admission occur earlier, making the indicated horse power less. On light loads the compression is often excessive, frequently attaining a value above that of the boiler steam. Should the load become heavier, the governor would so act on the valve gear that the events would take place later.

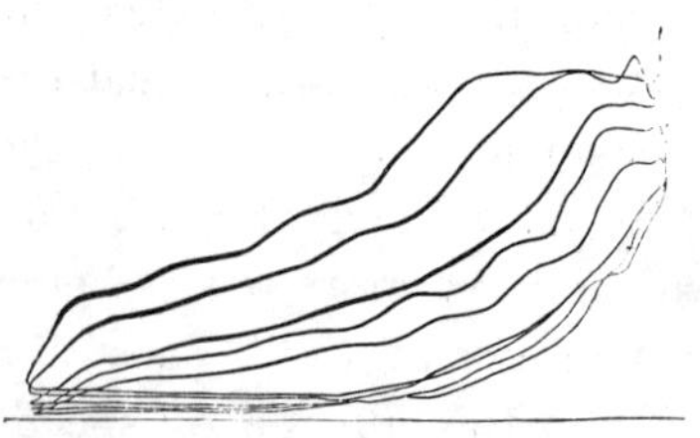

FIG. 222.—Effect of Automatic Cut Off.

In this method of governing, the steam supply is diminished at light load because the volume of the cylinder, filled at cut off, is decreased.

In both methods of governing the total quantity of steam is diminished as the load decreases, but the amount used in an hour per horse power increases. The **steam consumption,** as this amount is called, increases because of the effect of initial condensation at the earlier cut offs, or lower pressures.

The steam consumptions obtained with the two methods of governing, when applied to an experimental engine equipped for determining which of these methods was the better, show that the quantities do not differ greatly at the same power. The automatic cut off method of regulation is the one most prevalent in the United States.

Another point to be noted in regard to the steam consumption is that, although it diminishes with a later cut off, this only occurs up to a certain point, and beyond this it increases because steam is not expanded sufficiently after cut off. In these latter cases we are discharging into the exhaust much of the energy in the steam which could be utilized by expansion.

CHAPTER VIII.

GOVERNORS AND VALVES.

As many valves controlling the steam distribution of an engine are made in their particular forms because of their use in governing, and as special arrangement of governors are used with these valves, the valve and the governor will be considered together.

Governors. — **Plain D slide valve and throttle governor.**—In Chapters V and VI the action of the D slide valve was discussed. It was seen how the single valve controlled the admission and exhaust, and also that these events would always occur at the same point of the stroke, were the eccentric fast on the shaft. A **throttle governor** which may be used to regulate the speed, when such a valve gear is employed, is shown in Fig. 223. It consists of a valve made of two discs AA, connected to a spindle M, which is moved by the action of centrifugal force on two revolving balls. The discs AA, moving within an opening, are of such a fit that the valve may move freely, and yet without any play. The pressure above the upper disc, and below the lower one is the same, being that of the boiler steam, while that between the discs gives equal upward and downward pressures, no matter whether the valve be open wide or closed. Such a valve is said to be **balanced.** Valves of this form are sometimes used as throttle valves. The balls BB are attached to arms C on the sleeve D, which is held in the vertical bearing E. The bearing E is supported on a bracket which is cast to the top F of the valve casting, and carries a horizontal bearing G. To the lower end of the sleeve D is attached a **bevel gear** H, which engages with a gear

I on the horizontal shaft. This shaft is turned by a belt from the shaft of the engine through the small pulley K. In this way the sleeve D is turned, causing the balls to revolve about

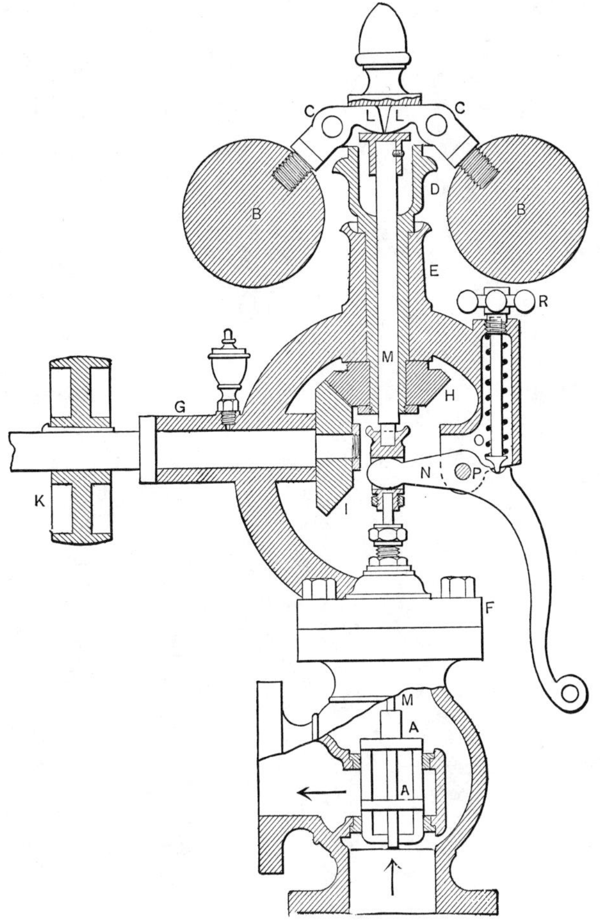

FIG. 223.—Throttling Governor.

the axis. As the speed increases, centrifugal force drives the balls further from the center, the balls rise and the inner end L of the arm C is depressed, moving the end of the spindle M downward. This partly closes the valve, throttling the steam and reducing the pressure on the engine side of the valve.

To regulate the speed at which the engine will run, a lever N and a spring O are used. This lever is pivoted at P, and the spring, the compression of which is regulated by the screw R, acts so that the inner end of N is forced upward. This end enters a slot on the spindle M, and keeps the valve open until the engine is running so fast that the pressure, due to the centrifugal force of the balls, overcomes the pressure due to the spring. This speed can be changed by the adjusting screw R. The spindle M is guided by the sleeve through which it can easily move.

Centrifugal balls, with a spring between to regulate their action, are sometimes mounted directly on the shaft of the engine, and are attached by a collar and sleeve to a system of levers and rods which act on a balanced throttle valve, thus accomplishing the same result as that previously described.

Watt governor.—Another form of governor, known as the Watt governor, is shown in Fig. 224. It is the earliest form of governor invented, being originally applied to a throttle valve and is now often used for Corliss engine governing. It consists of the two heavy balls AA, carried by the arms BB, which are pivoted on the end of spindle C. These arms are attached, by the rods DD, to the sleeve E, which contains a groove in which is the ring F. When the spindle C, which is supported and guided by the column G, is rotated by the gears and pulley beyond a certain speed, the centrifugal force of the balls will begin to raise the sleeve. If now the ring is attached to a throttle valve, the action is that before described.

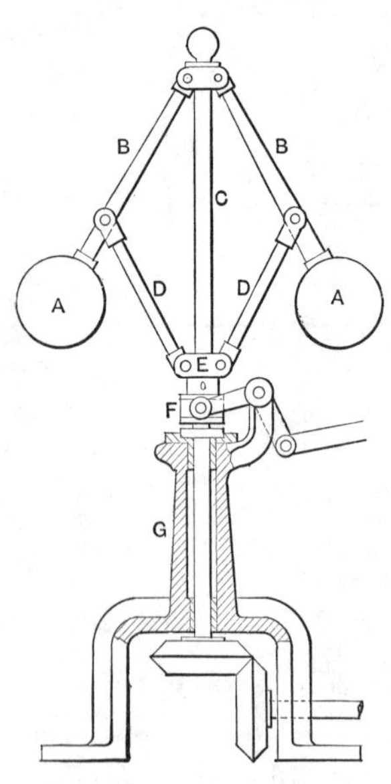

FIG. 224.—Watt Governor.

Porter-Allen governor.—As the Watt governor required

quite a change of speed to raise the governor from the lowest to the highest points, another form was introduced by Mr. C. T. Porter in 1862. This governor, shown in Fig. 225, is run

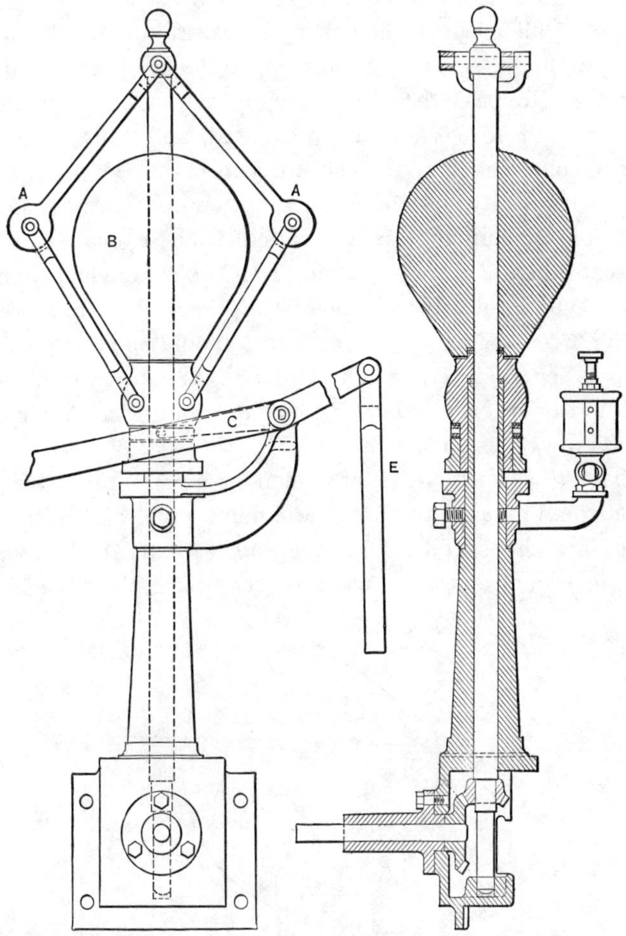

FIG. 225.—Porter-Allen Governor.

at a high speed. The centrifugal balls AA are quite small, and act on a sleeve which carries a sliding weight B of considerable size. The sleeve is also attached to the lever C, which is pivoted at D. The rod E, extending from this lever, is attached, in the Porter-Allen engine, to a link block which is connected to the end of a radius-rod. A **dash-pot** is attached to the lever. This consists of a cylinder, in which is a piston fitting snugly, but containing one or more small holes. The cylinder is filled with oil, which passes, from one side of the piston to the other, through the small holes. As the oil requires time to pass through these openings, any momentary changes of the governor are prevented, and thus a steadier action is obtained. If a dash-pot were not used, the slight changes of speed would keep the governor constantly changing in position, or **hunting,** as it is called. These changes in the position of the governor, occurring after the engine had changed speed, would probably aggravate the change so that greater variation would occur. All governors should be equipped with dash-pots or their equivalent, allowing a steady change of speed to move the governor, but preventing a momentary fluctuation to cause any change.

Shaft governors.—As has been noted, a Stevenson link motion could be used to change the point of cut off and thereby to regulate the speed of an engine with a D slide valve. A single eccentric can be used to accomplish the same object

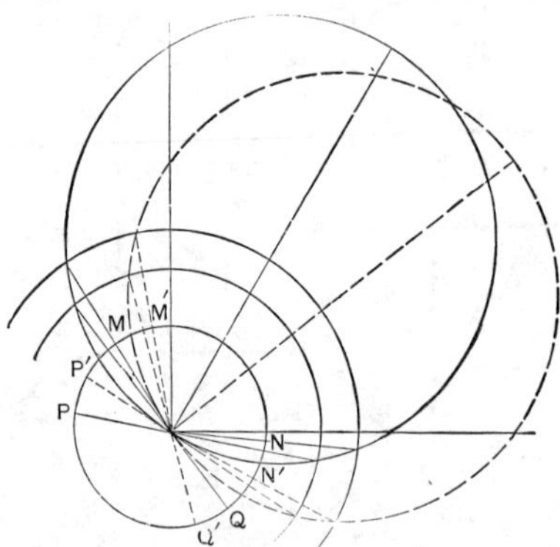

FIG. 226.—Governing by Shifting Eccentric.

if it is moved on the shaft by some means. If the eccentricity remains constant and the eccentric is moved, the events would all occur earlier, as seen in Fig. 226, but release and admission would be changed by the same amount as cut off and compression, and this would give a distribution of little value. In the Stephenson link motion it was seen that the virtual eccentricity as well as the angle of advance is altered, and that this method gave distributions which were satisfactory. To do this with a single eccentric, the eccentric must be moved so that the eccentricity changes with the angle of advance. Fig. 227

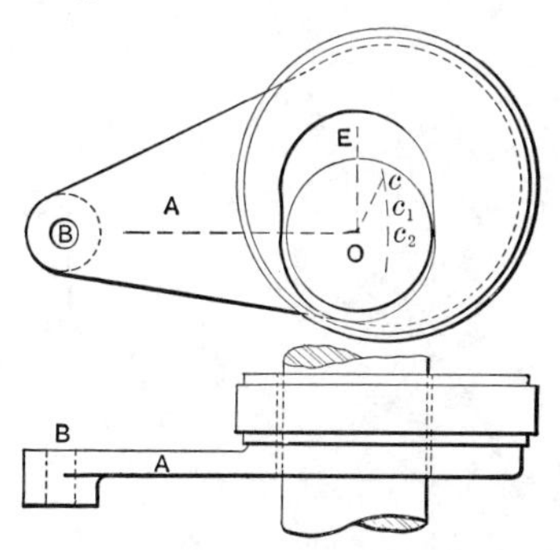

FIG. 227.—Automatic Cut-off Governor.

shows one method of moving an eccentric to accomplish this. The sheave of the eccentric is mounted on an arm A, which is pivoted on one of the arms of the fly wheel at B. This sheave contains a slot which passes over the shaft, whose center is O. c is the center of the eccentric, cO being the eccentricity, and cOE the angle of advance, the governing arrangement being attached to the sheave. To move it so that c has the position c_1, c_2, it is seen that the angle of advance increases and the eccentricity becomes less in the same manner as the virtual eccentric of a Stephenson link with open rods. The governing apparatus is pivoted on the arms of the fly wheel and attached to this sheave by rods and regulated by springs so that the engine must reach a definite speed before the eccentric is moved.

Rites Inertia Governor.—A modern form of shaft governor is shown in Fig. 228. This governor is usually placed at the end of the shaft where the fly wheel is overhung. It consists of the inertia weight A pivoted on one of the fly-wheel arms at B, and containing a pin C, to which the eccentric rod is attached. The regulating spring D pulls the weight A against the stop E on the fly wheel. As the speed increases, the engine turning as shown by the arrow K, the weight tends to move

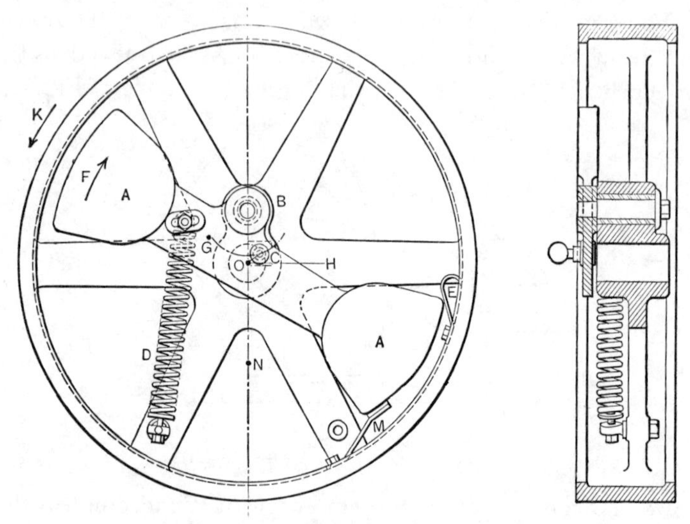

FIG. 228.—Rites Inertia Governor.

in the direction shown by the arrow F, because of the centrifugal force, the center of gravity G not being coincident with the point B. As the motion of the engine becomes greater, this centrifugal force overcomes the spring tension, and the weight moves. In the first position, since C is the center of the eccentric pin, and O is the center of the shaft, CO is the eccentricity and COH is the angle of advance, N being the center of the crank; and, as the speed changes, the length CO becomes smaller, while COH becomes greater in a manner equivalent to a Stephenson link with crossed rods. A drawing of the valve diagrams for different positions of C will show how the different events of the stroke change as the weight A moves. There are two other actions in this governor which are important, and which give the value to this special type. They are the inertia effects. When the engine is moving at a definite speed and the load is suddenly removed, causing the engine to increase in speed, the inertia of the arm causes it to lag behind the fly wheel, the speed of which is increased, and hence the motion of the weight relative to the fly wheel is as shown by the arrow F, the engine turning in the direction K. The increase in the speed of the engine also tends to move the center of gravity of the weight faster, but inertia causes this to lag, and hence this also tends to turn the weight in the direction of F.

The spring D, which acts on the weight, is so proportioned and located that as the weight moves its force so varies that the number of revolutions are almost constant at different loads. The spring M prevents the movements under the small fluctuations of speed during one revolution.

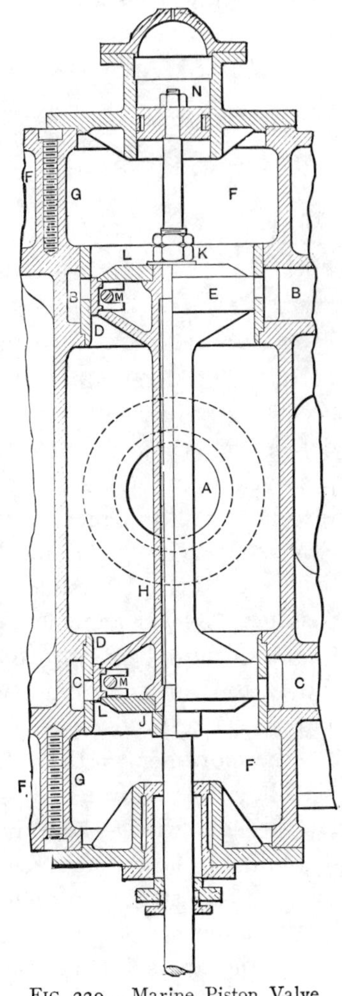

FIG. 229.—Marine Piston Valve.

***Valves.*—Piston Valves.**—This valve was designed primarily on account of the impossibility of making large slide valves effective because of their want of tightness. As piston valves decrease the friction and increase the port opening for the same space occupied by the valve, they are used when large valves must be used, with anything but very low pressures. The form shown in Fig. 229 is that used in the United States Navy and is in reality a D slide valve wrapped into the form of a cylinder. Steam is taken inside, entering through the pipe at A. It enters the head end of the cylinder through the passage B and from the crank end is exhausted through the passage C. The ports are made in liners D (shown in Fig. 230), which are forced into bored holes in the cylinder casting after being finished all over. These ports are really diamond-shaped holes which extend around the circumference, the metal strips, which form the division between them, holding the two parts of the liners together and preventing the edges of the piston rings E, which are used on these valves, from springing into the ports. This liner allows the passages to be made easily and accurately, as a small displacement of the opening in the cylinder casting is adjusted by the position of the liner. The steam passages B and C extend completely around the valve, and it is to be noted how much larger they are than the ports. The exhaust is taken off from the spaces FF, the section shown being cut through a stiffening bolt GG in the center of the passage. The arrangement of the lap is peculiar in this engine, the steam laps being $\frac{29}{32}''$ on the top or head end, and $\frac{27}{32}''$ on the bottom or crank end, while the exhaust laps are $-\frac{5}{16}''$ and $-\frac{1}{16}''$ respectively. The eccentricity in this case being $2''$, the movement of the valve is $4''$.

FIG. 230.—Valve Liner.

The valve shown in Fig. 229 is made of a central casting H, through which the valve rod extends and to which, by the collar J and the nuts K, two plates LL are held. The piston ring is a single ring regulated by a bolt M on the inside. The stuffing box is shown at the bottom of the steam chest and at the top is the **balancing piston** N. This piston is exposed on its lower surface to the exhaust steam pressure and above to the atmosphere. This balances the pressure due to the area of the valve rod passing through the stuffing box, and is also made enough larger to support part of the weight of the valve, thus relieving the pressure on the eccentrics.

Double-ported Valves.—Another method of increasing

the amount of port area for small movement of the valve is by the use of a double-ported valve. This valve was spoken of when describing the Weston engine, Fig. 112, but the form there shown is intended to act as a double-ported valve only at the points of admission and cut off. For marine engines a valve has been used in which there are two admission areas which are active throughout the entire movement of the valve. Figs. 231 and 232 show different views of the valve. The steam surrounds the valve and enters the passages AA′, which extend from one side of the valve to the other. When the valve moves upward steam can enter the two ports B, C, while exhaust occurs at E and F. The exhaust steam which enters the upper part of the valve passes over the top of the passage A′ to the central portion, and thence to the exhaust port G. To reduce the friction load on this valve, a packing ring H, on the back of the valve moves in contact with the **pressure plate** K, which is part of the valve cover. The back of this valve is made of the same area as the face, to balance the pressure. These valves, which are usually very large (the valve shown being about 24″ × 36″)

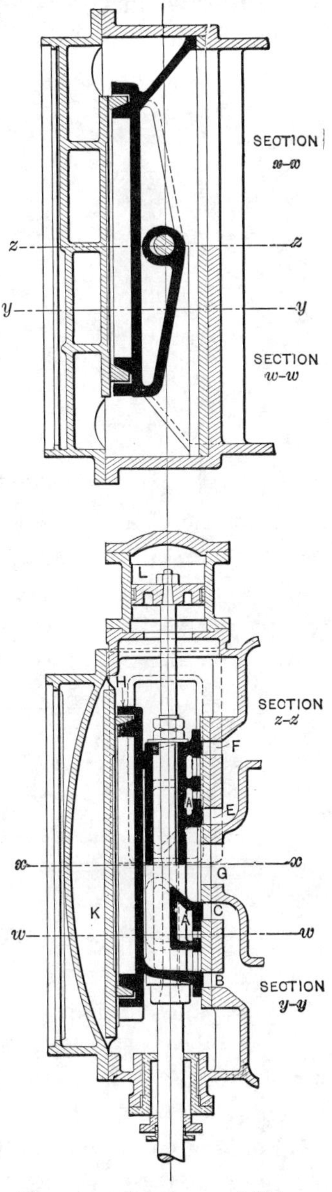

FIG. 231.—Double-ported Valve.

are so heavy that it is advisable to balance their weight, as shown by the use of the auxiliary piston L. Fig. 232 shows

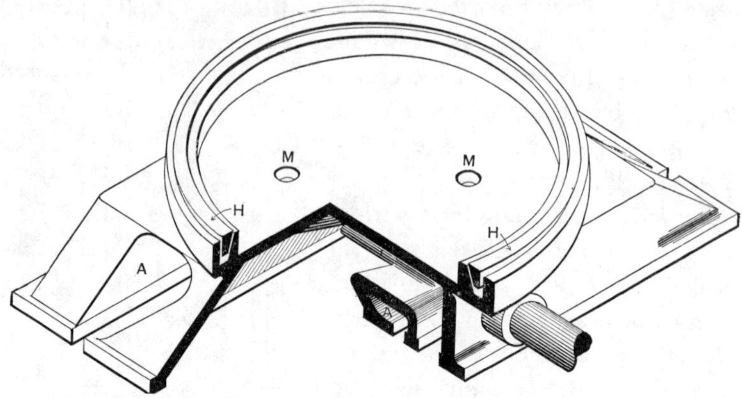

FIG. 232.—Double-ported Valve.

an isometric view and section of this valve. The holes shown at M are to allow leakage past the ring to escape into the exhaust.

Double-ported piston valve.—Fig. 233 shows the piston valve from an Armington and Sims engine. In this valve the

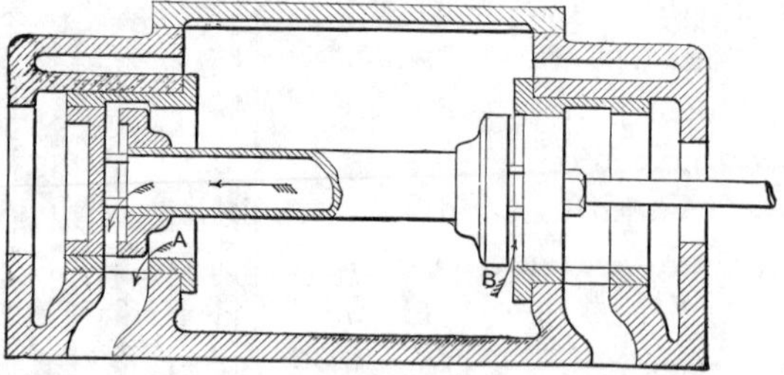

FIG. 233.—Armington and Sims Valve.

steam is taken inside, steam entering the left-hand end by passing the corner A, and also by the corner B through the center of the valve. For the exhaust events the valve is single-ported.

Four-valve engine.—The single-valve engine necessitates a large clearance and also the use of the same passage for live steam after it has been cooled by the exhaust steam. To overcome these objections four-valve engines have been introduced, of which the first successful one was that made by Mr. Geo. H. Corliss. In this engine two valves were used for steam and two for exhaust, which being placed near the ends of the cylinder, required short steam passages, thus reducing the amount of clearance. This arrangement was so successful that the method has been copied by many engine builders who have modified the details of the valves and the driving mechanism. This to-day is the gear mostly used for large stationary engines. The valve motion of the Hamilton-**Corliss engine** is shown in Figs. 234 to 237. Steam enters a **box** A, Fig. 234,

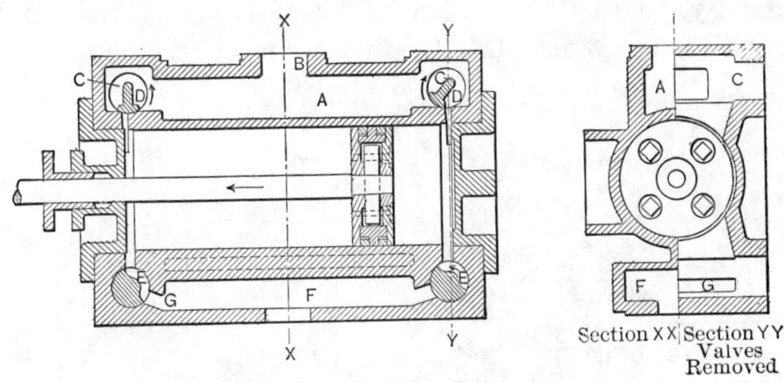

FIG. 234.—Hamilton-Corliss Cylinder.

on the top of the cylinder at the point B; it there divides, passing to the valve boxes CC at each end. These valve boxes contain the **steam valves** D, shown in detail in Fig. 235.

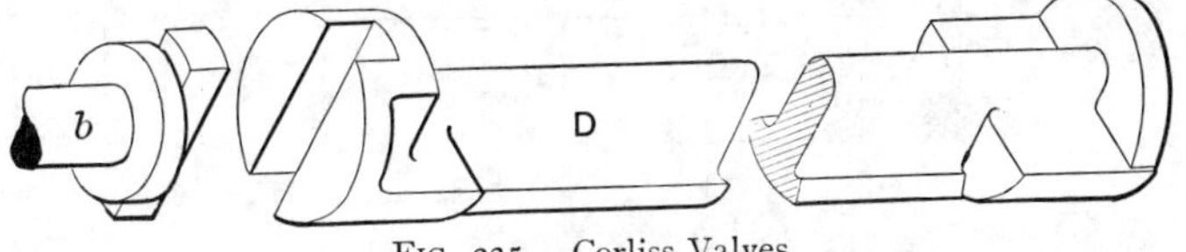

FIG. 235.—Corliss Valves.

These are bars of cast iron having cylindrical ends and a thin portion connecting the ends. The valve face is a continuation of part of the cylindrical surface with a web on its back to stiffen it. The **exhaust valves** EE, on the lower side of the cylinder, are similarly constructed, with the exception that the length of the arc of the connecting portion is greater than that of D. These valves E lead into the exhaust box F, through the exhaust ports G.

If the valve D be rotated from its middle position in the direction of the arrow, so that the surface is moved an amount equal to the lap, steam will be admitted, and when the valve is moved quickly back the cut off will take place sharply, giving a decided corner to the indicator diagram in place of a rounded one. After this, the steam expands to a point near the end of the stroke when release takes place through the lower port. The manner in which these valves are operated is shown in Figs. 236 and 237. The eccentric rod H is attached at the point K to the **wrist plate** L which turns on M. This oscilla-

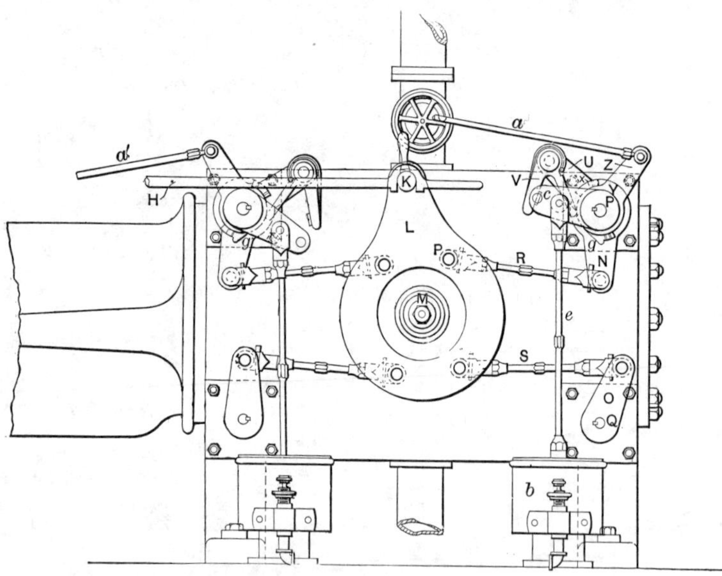

FIG. 236.—Corliss Valve Mechanism.

tion of the wrist plate moves the arms N and O, pivoted at P and Q by the rods R and S. To the shaft Q, to which O is keyed, is attached the exhaust valve E of Fig. 234, so that it is oscillated as the eccentric revolves, opening and closing the exhaust at the proper time. The arm N is not directly attached to the steam valve, but to a sleeve T, Figs. 236 and 237, which also carries the arm U, and on the end of U is pivoted the hook V. V is made with two arms, one of which carries the catch W, and the other travels over a cam X. The cam X is mounted on a collar Y, which turns on the same bearing as that on which the sleeve T moves. The arm Z, on Y, is attached to the governor by the rod *a*, so that its posi-

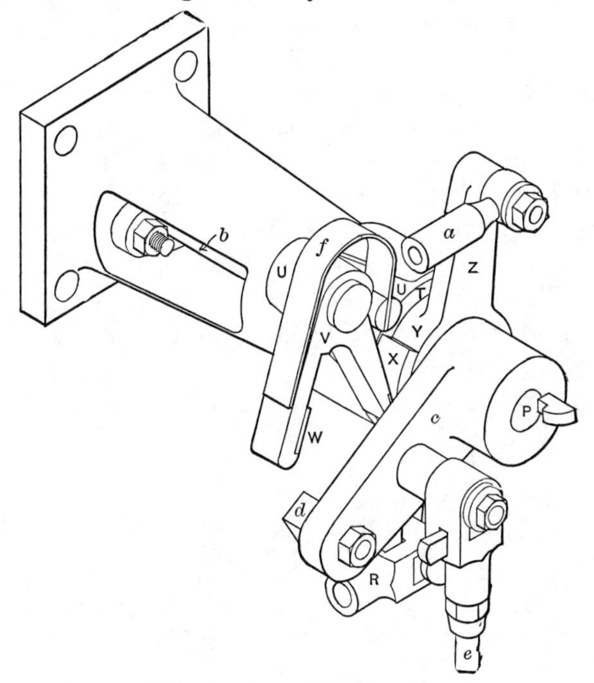

FIG. 237.—Corliss Releasing Gear.

tion is determined by the speed of the engine. To the valve spindle P is keyed an arm *c*, on the side of which a block *d* engages with the catch W. To this arm is also attached the vacuum dash pot *b* through the rod *e*. If the wrist plate moves so that N is moved to the left, the arm U is raised, and with it, by means of the hook V, the arm *c*. This opens the steam valve. When, however, the inner arm of the hook V strikes the cam X, the catch W is pulled away from the block *d* and the vacuum in the dash pot pulls the arm *c* downward, closing the valve sharply. As the load on the engine changes, the cam collar Y is moved by the governor so that the cam X strikes the arm of the hook either earlier or later in the stroke.

The spring *f* forces the hook against the block so that it will catch at the proper time. In case the belt which usually drives the governor of a Corliss engine should break, and thus allow the maximum quantity of steam to enter the engine, as the balls would then fall to their lowest position, a safety-cam *g*, Fig. 236, is placed on the cam collar. When the governor is in its lowest position the arm Z is moved over so far that the inner arm of V travels over the cam *g* when U is in its lowest position. This moves the catch so that it cannot engage with the block on *c*, and hence the valve will not open and the engine will stop. To permit one to start the engine, a safety pin is placed in the governor column so that the governor cannot reach the position at which the hook will not catch. This safety pin should always be removed when the engine is running, or the breaking of the belt would cause the engine to run away.

Fig. 238 shows the construction of a vacuum **dash pot**.

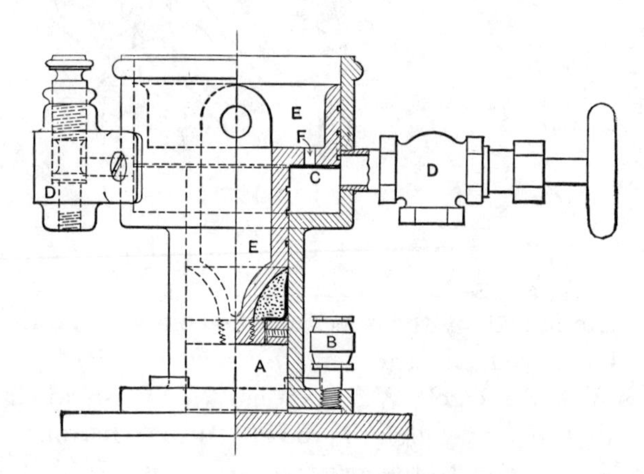

FIG. 238.—Vacuum Dash Pot.

The lower cylinder A is connected to the atmosphere by the check valve B opening outward, while the upper cylinder C is open through the partially closed valves DD and the inwardly opening leather check-valve at F. If the double piston EE is drawn upward, a vacuum is formed in the lower cylinder, while air slowly enters the upper one. On the removal of the catch W from beneath the block *d*, on the arm *c*, Fig. 237, this vacuum draws the piston downward and with it the arm *c*. This takes place rapidly at first, as there is a vacuum in the lower cylinder, but as the pistons approach the end of their travel the air in the upper cylinder is being driven out through the small openings in DD, and as this cannot take place fast enough, the air is compressed and the piston brought to rest. The rapidity of the fall can be regulated by adjusting the valves DD.

In the engine shown, only one eccentric is used to drive both the steam and the exhaust valves, but as we cannot always secure sufficient range of cut off or proper steam distribution, separate eccentrics and wrist plates are used for the steam and exhaust. These are usually found on large engines.

Porter-Allen valves.—Another four-valve engine is the Porter-Allen engine, invented by Messrs. Charles T. Porter and John F. Allen. Fig. 239 shows the construction of

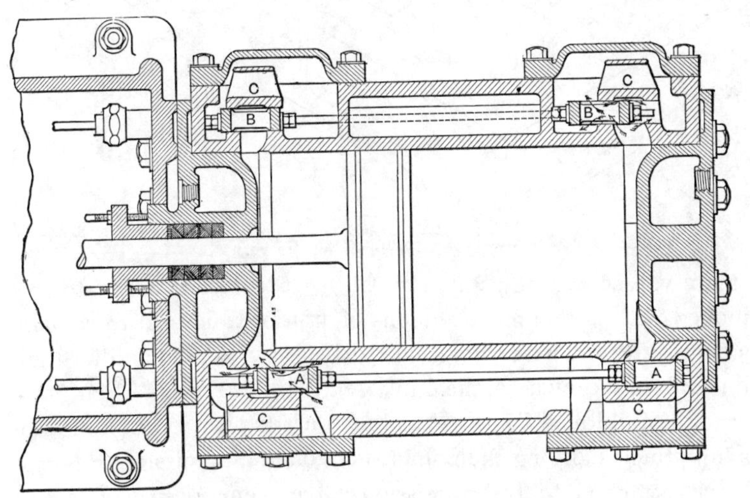

FIG. 239.—Porter-Allen Cylinder.

the cylinder of this engine. Steam valves A are mounted on two separate valve stems, the stem of the head valve passing through holes in the crank valve. The exhaust valves B are attached to one stem. The valves are made of cast iron in the form of rectangular frames, as shown in Fig. 240. They are double-ported valves, as is clearly seen in the figure. The pressure plates C on the back of the valves prevent excessive friction, and in case of excessive compression the plates over the steam valves allow the valves to be raised from their seats. The exhaust valve is placed at a lower level on the side of the engine to permit of draining. The pressure plates over the

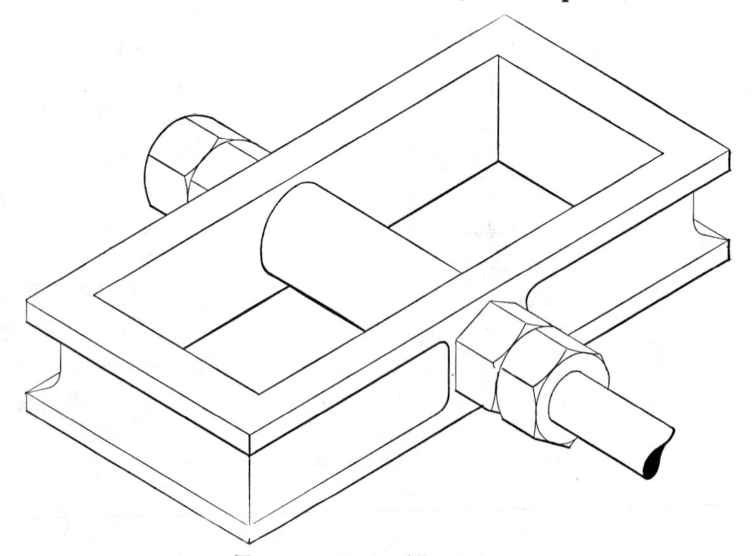

FIG. 240.—Porter-Allen Valve.

steam valves are adjustable to allow the steam valve to be moved with a minimum amount of friction and still to remain steam tight. These valves are driven by a modified form of link motion known as the Fink gear. The Porter-Allen governor is employed to move a block on a radius rod in a slot in a link, thus changing the points of cut off and admission, this rod being attached to the admission valve stems only. The exhaust valves receive their motion from a fixed point in the link.

The Meyer expansion valve.—To vary the point of cut off an additional valve is sometimes used moving on the back of the main valve, and is known as an **expansion valve.** The Meyer expansion valve, Fig. 241, is of this form, and consists of two blocks AA, which are attached to a threaded valve stem B. The main valve C has pieces added at the ends, so that there are two steam passages EE formed in the end of the main valve. The exhaust cavity F is similar to that ordinarily found with the D slide valve, causing release and compression at fixed points. The blocks AA are driven by a separate eccentric, and when a passage E is covered by one of the blocks, steam

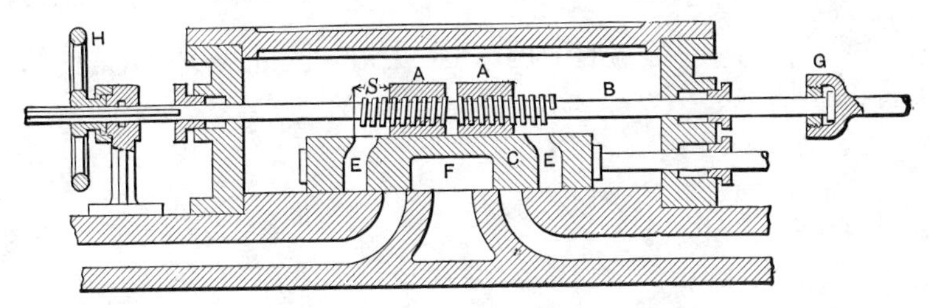

FIG. 241.—Meyer Valve.

is cut off from that end, no matter what the position of the main valve may be. The auxiliary valve stem is so attached to its eccentric rod by a cap G that it may be turned on its center line by means of a hand wheel H, on the outer end, through which the tail rod extension passes. The threads for the two blocks are right- and left-handed, and turning the rod in one direction will cause the blocks to separate, producing an earlier cut off as the passage E is closed sooner. If the wheel H is turned in the opposite direction, the blocks are brought closer together and cut off will take place later. The admission is governed by the main valve, which is known as the **distributing valve,** while the blocks form the expansion valve.

Valve diagram for Meyer valve.—It has been shown that the motion of a valve driven by an eccentric is represented by the equation

$$x = r \sin (\omega + \delta).$$

If now another valve is driven by a separate eccentric of radius r' and angle of advance δ', its motion relative to the cylinder will be

$$x' = r' \sin(\omega + \delta'),$$

and its motion relative to the other valve will be

$$X'' = x - x' = r \sin(\delta + \omega) - r' \sin(\delta' + \omega).$$

This can be reduced to the expression

$$X'' = (r \cos \delta - r' \cos \delta') \sin \omega + (r \sin \delta - r' \sin \delta') \cos \omega,$$

which is also represented by a circle the diameter of which is found as shown in Fig. 242.

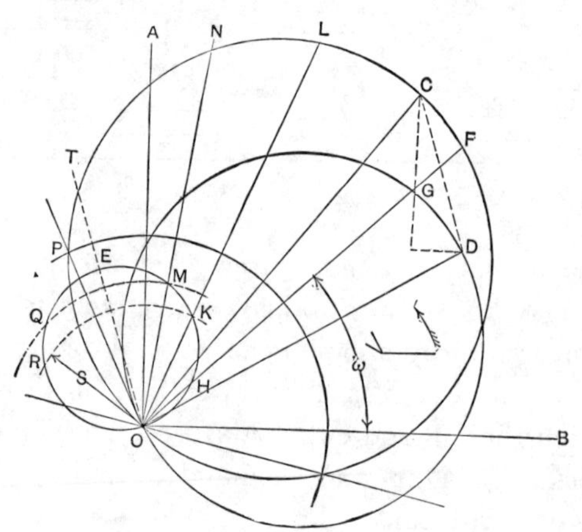

FIG. 242.—Meyer Valve Diagram.

From O the point of intersection of OA and OB, two perpendicular lines, lay off OC equal to r and OD equal to r', AOC equal to δ and AOD equal to δ'. If from O, OE is laid off equal and parallel to DC, it will be the diameter of a circle which represents the movement of the expansion valve relative to the main valve, r and δ referring to the main eccentric and r' and δ' to the expansion valve.

Suppose the crank has moved through an angle ω from its dead-point. The main valve has moved from the mid-position a distance OF and the auxiliary valve a distance OG, the relative motion being the difference GF, which is also shown in the relative circle by OH. When this relative motion amounts to the distance s, Fig. 241, cut off will occur, hence if OK, Fig. 242, represents this distance, L is the position of cut off. If by the hand wheel H, Fig. 241, the blocks are brought closer together until s equals the distance OM, Fig. 242, cut off will occur at N. In this way the cut off can be varied, the latest cut off by the auxiliary valve occurring at T; but in this case the valve would immediately open and cut off would occur by the main valve at P, where the steam lap OP cuts the valve circle of the main valve. Where the "s circle" cuts the auxiliary circle a second time, as at Q and R, is the location of the point at which the expansion valve uncovers the port in the main valve, and this should not occur before the main valve cuts off (at P) since it is not desired to have a second admission of steam.

CHAPTER IX.

CONDENSERS AND MULTIPLE-EXPANSION ENGINES.

THE early piston engines employed the condensation of the steam in conjunction with the pressure of the atmosphere to drive the piston. In these engines, used for pumping water in the English mines, the cylinder was connected to a boiler in which steam was made at about atmospheric pressure, and, as the piston was drawn upward by the weight of the pump rods acting through a walking beam, this steam flowed into the cylinder. When the piston reached the top of its stroke, the communication to the boiler was shut off and water was introduced into the cylinder. This abstracted the heat from the steam and condensed it, forming a partial vacuum, and, as the upper end of the cylinder was open to the atmosphere, the air pressure forced the piston inward, lifting the plunger of the pump. At each downward stroke the cylinder was cooled, causing a great loss of heat, and, to improve the engine, Watt, in his earlier engines, condensed the steam in another vessel, called the condenser, and later introduced the double-acting principle. Condensers are now used to reduce the back pressure against which the piston must act, the increase in the work done at the same cut off being shown in the diagram, Fig. 243, taken from a 10″ by 24″ engine. In the first case the area of the indicator card is 5.10 square inches, the back pressure being 17 pounds absolute, while in the second one the area is 6.50 square inches and the back pressure 3 pounds absolute. Thus, by condensing the steam, we gain 27 per cent. in the power developed at the same cut off. Although the actual quantity of steam may be slightly increased, on account of greater initial condensation when a condenser is used, the quantity of steam per horse-power hour is greatly diminished. There are two methods of condensing the steam; one, by bringing the steam in contact with a jet of water, and the other, in which the steam strikes against a metallic surface cooled by water, the apparatus in which these actions take place being called jet condensers and surface condensers respectively. In each of these forms

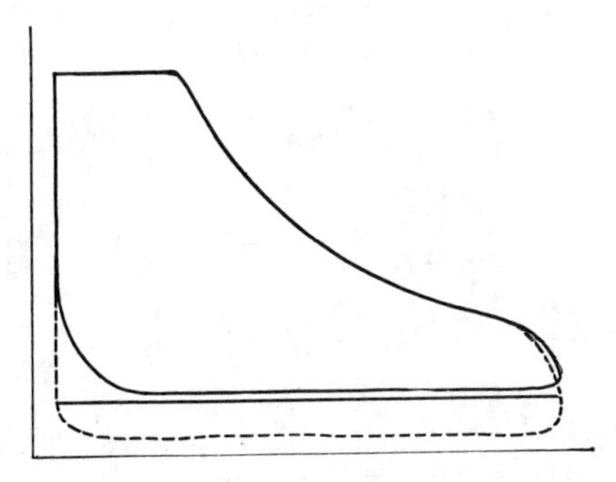

FIG. 243.—Effect of Vacuum.

water must be pumped out from the vessel containing the condensed steam against atmospheric pressure, and, as air is present in most steam and in all water, the pump to perform the operation has to be larger than it would be were it necessary to remove the water alone. This pump is usually termed the **air pump.** Fig. 244 shows the construction of a jet condenser attached to an air pump. Steam enters at A, passes through the deflecting cone valve B, and mixes with the condensing water which is drawn through the **suction injection** C by the vacuum in the **condensing chamber** D. The cone valve E regulates the supply of condensing water and discharges against the **spray plate** F, which deflects it into the condensing chamber, where it mixes with the steam which has passed above the spray plate. The water resulting from this thorough mixture of steam and water, together with the air liberated, is removed by the air pump G through the passage H and the valves J, as the pump piston moves to the left. On the return stroke the mixture is driven into the discharge L through the valves K. If a small amount of steam enters the condenser, water may be decreased by screwing down the cone by means of the hand wheel and nut M, and the speed of the pump may be decreased. If the amount of steam becomes

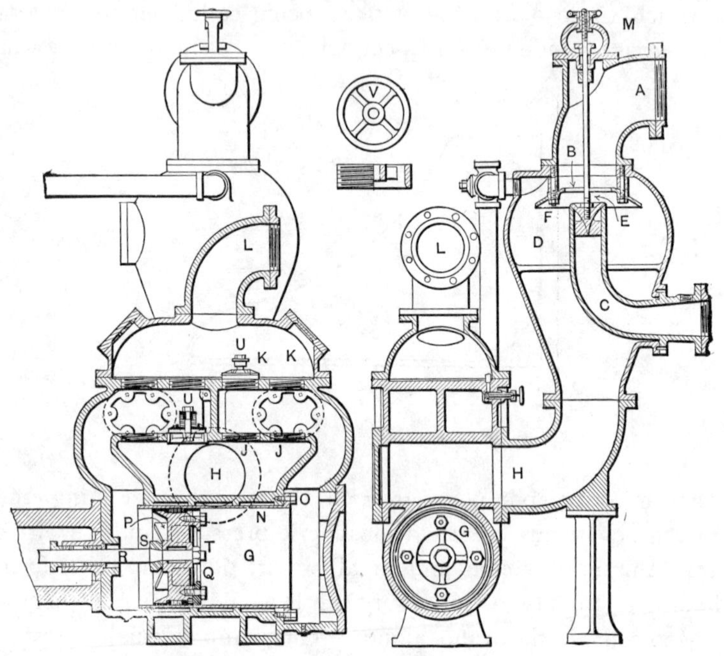

FIG. 244.—Snow Jet Condenser and Air Pump.

greater, this small amount of water would not condense all the steam, and hence the vacuum would be impaired and, in such a case, raising the cone would restore the proper conditions.

The cylinder of the air-pump from the jet-condenser shown contains a brass liner N, which is held in place by the flange O. The piston consists of a cast iron body P, with two liners shown in black, held in place by the **follower** Q. Between the liners is the packing of canvas indurated with rubber. The **pump rod** R is fastened to the body of the piston by the collar S and the nut T. The valves at J and K, which are called **disc valves,** are made of vulcanized india-rubber or sheet metal. If made of india-rubber they must contain certain ingredients to resist the heat, and also the mineral oil which is present in the exhaust steam from the engine. The constituents of this material, according to Sennett and Oram, are 70 per cent. oxide of zinc, $1\frac{3}{4}$ per cent. sulphur, and the remainder the best caoutchouc. The valves are free to move vertically on the valve stems, being forced on their seats by washers and springs, nuts holding their guard in place as shown at U. The valve seat V is made of brass, and is screwed into the cast iron partition as shown at J and K.

Pump valves.—Other forms of pump valves are the flap or clack valves, the wing or direct-lift valves, ball valves, and double-beat valves. The **clack valve,** Fig. 245, consists of a piece of leather A, held beneath a plate of metal B, and having a stiffening plate C fastened under it. These plates prevent the pressure from forcing the leather through the valve-seat,

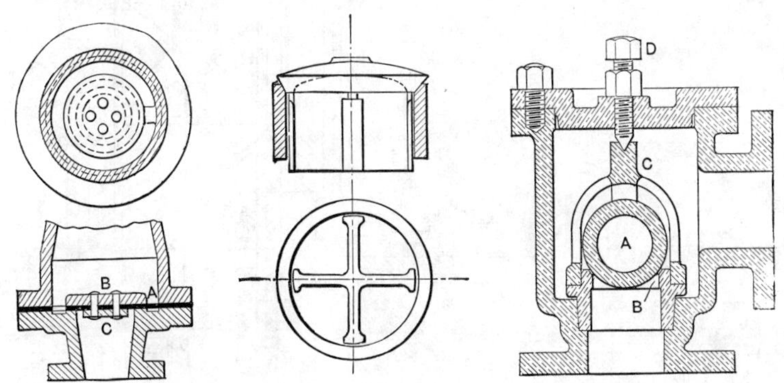

FIG. 245.—Clack Valve. FIG. 246.—Lift Valve. FIG. 247.—Ball Valve.

the leather acting as a hinge and making a tight joint.

The **direct-lift valve** is made as shown in Fig. 246. The disc of metal has a conical face which fits a conical seat, the surfaces having been ground together. The valve is guided by wings or by a central spindle which passes through a guide fastened to the body of the seat.

A **ball valve** from Low & Bevis' machine design is shown in Fig. 247. A spherical ball A rests against a seat B, and is held in place by a cage C, which is secured by a set screw D. Ball valves are used in artesian well pumps. A **double-beat valve** is made with two discs united, as shown in the throttle governor, Fig. 223, each disc having its independent conical seat.

Weiss Condenser.—A form of jet condenser has been invented by F. J. Weiss, of Basle, in which the air pump is used

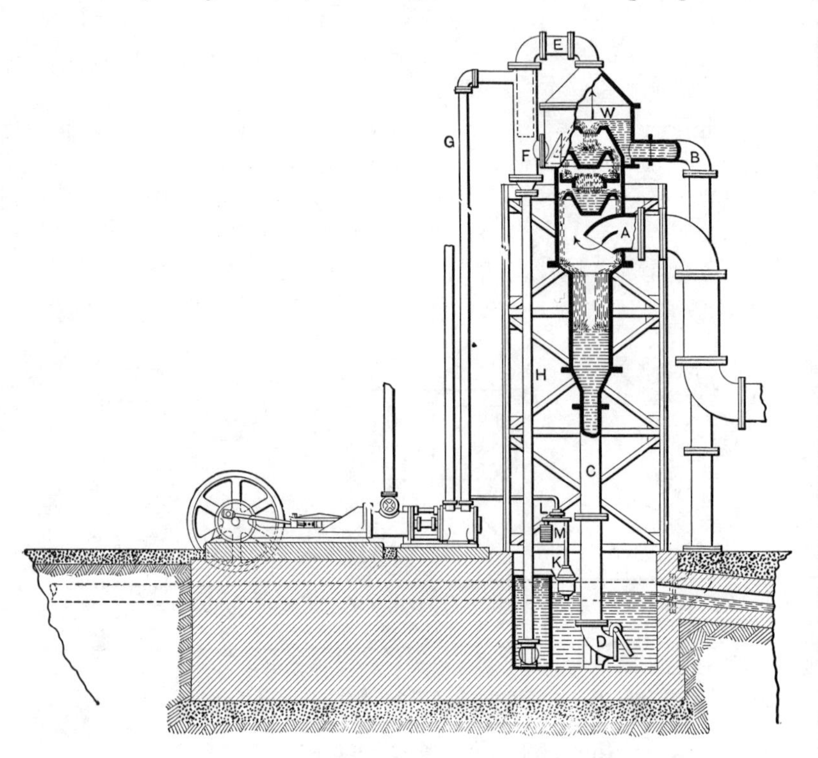

FIG. 248.—Weiss Condenser.

only to remove the air and some little vapor. It is known that, if the air be exhausted from a vertical pipe, the lower end of which is beneath a water surface, water will rise to a height of about thirty-four feet. Any additional water entering at the top of this column would force out an equal amount at the bottom. Fig. 248 shows the construction of the **Weiss**

counter-current condenser.* Steam enters at A, while the condensing water enters at B. This water is drawn up part way by the vacuum in the **barometric tube** C, a circulating pump forcing it the remainder of the distance. The water falls from W over plates, forming a cascade, and it here mixes with the steam, which tends to move upward as the pressure is lower at the top of the tube C, since the air pump is attached at this point. The condensed steam and condensing water fall by gravity through the barometric tube C, and out at the bottom through the check valve D. The air pump is connected with the top of the condenser through the pipe E, the separator F, and the pipe G. By means of this separator, which also has a barometric tube H to relieve it of water, the air is made so dry that the size of the air pump is small. This condenser makes it possible to use **injection water,** as we usually term the cooling water, at higher temperatures than are customary with other forms, and the apparatus is quite economical and simple.

In operation, when the quantity of water pouring through the separator F becomes large, the condenser acts as a parallel-flow condenser, rather than a counter-current one, in which the steam and water move in opposite directions. This parallel flow would soon clog the discharge pipe H and wreck the air pump. To restore the action to that of a counter-current one, the discharge from H passes into a bucket K, normally allowing the water it receives to pass through it. When the water comes down H faster than the holes in the bucket can care for, the bucket fills, and, dropping down, opens a valve L, allowing air to pass into the air pump and breaking the vacuum sufficiently to cause the condenser to return to its original method of working. A counter weight M closes the valve L as soon as the bucket is empty.

Surface Condensers.—When the condensed steam is to be used again for boiler feed because of the scarcity of pure water, as, for instance, on ocean vessels, the surface condenser is used.

---

* U. S. Patent Reissue 11591.

In this, any cool water is circulated through a series of tubes on the outside of which is the exhaust steam. The tubes are usu-

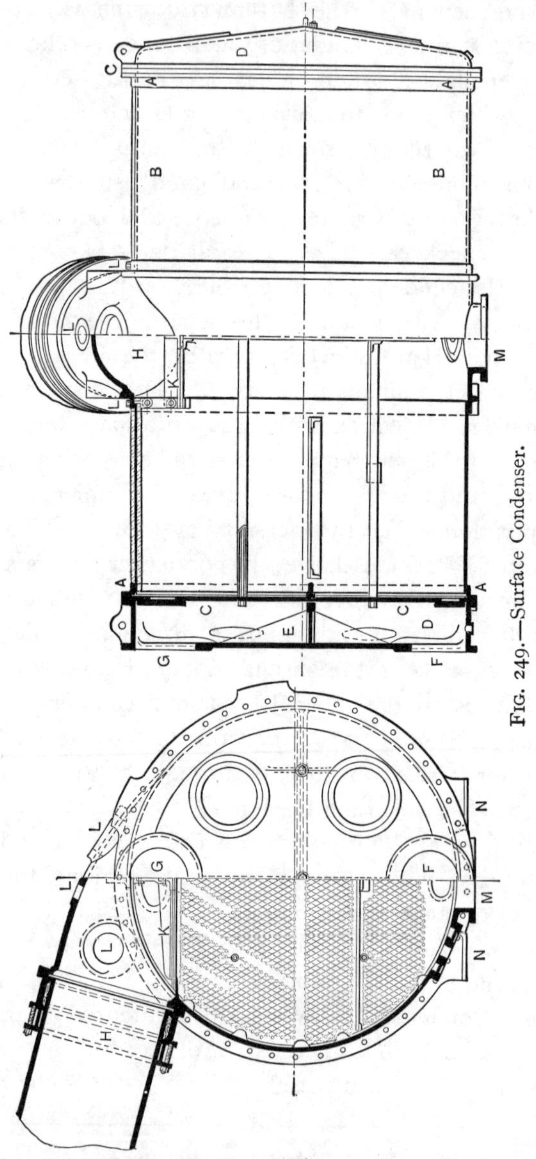

FIG. 249.—Surface Condenser.

ally made of thin brass, and readily conduct the heat from the steam to the water and thus condense the steam. The construction of a surface condenser is shown in Figs. 249, 250, and 251.

To flanges A, on a cylindrical shell B, are fastened the two **tube plates** CC, which are drilled to receive the tubes extending from one plate to the other. Outside of these plates are bolted the heads DD, the left-hand one of which has a web E fitting snugly against the tube plate, and thus forming two compartments. Water enters at the point F and passes through

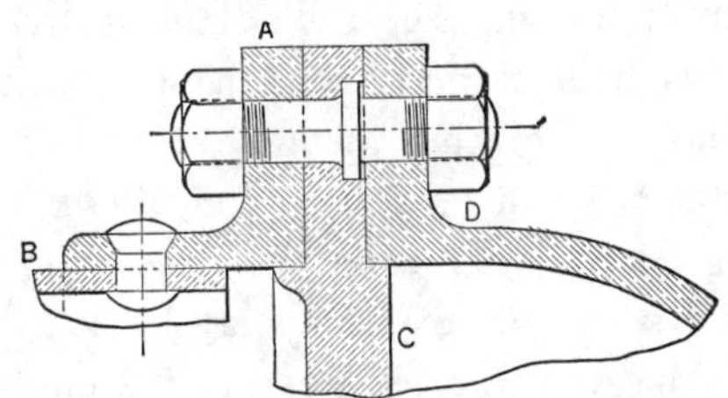

FIG. 250.—Surface Condenser Tube Sheet.

the lower set of tubes, which, in the condenser shown, consists of 1896 tubes five-eighths inch in diameter. At the right-hand end the circulating water ascends and passes through the upper set of tubes, 1879 in number, to the point G, where it discharges into a pipe leading to the sea. Steam from the main engine enters at H, through an expansion joint, and striking on a **baffle-plate** K, is distributed over the whole length of con-

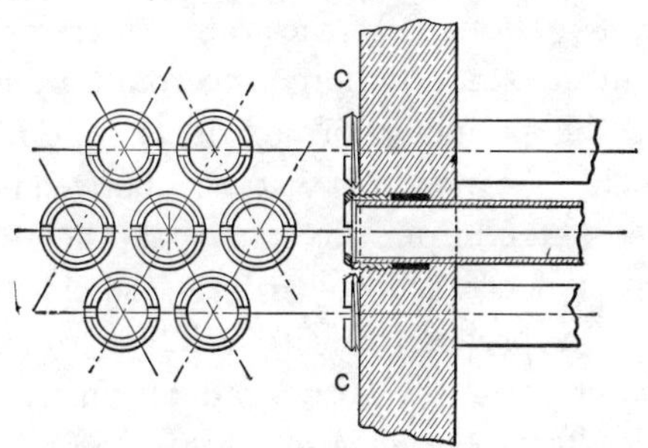

FIG. 251.—Condenser Tubes.

denser. The small opening at L is for the exhaust of the auxiliary machines, such as feed pumps, condensers, and circulating pump dynamo engines, etc. The condensed steam and air are drawn into the air-pump suction pipe from the bottom of the condenser at M. The feet NN serve to support the shell. The tubes are held in place by packing, as shown in Fig. 251. A hole is bored in the tube plate the size of the tube, and this is counter bored from the outside to a greater diameter, forming a stuffing box. This counter bore is threaded, and in it screws a brass **ferrule.** This ferrule presses a woven packing against the tube. The ferrule has a slot end so that it may be screwed up by means of a screw driver, and is bent over so that the tube is prevented from slipping out. The tubes are made of thin brass, which has been tinned on both surfaces to better resist corrosion. They are in this instance nine feet in length and each tube contains about $1\frac{1}{2}$ square feet of condensing surface.

Circulating Pump.—To drive the circulating water through the condenser, a pump of large size may be attached by a rocking lever and links to the cross head of the engine; but as the condenser is apt to become heated from the auxiliaries when the engine is stopped, the circulating pump is more often made independent of the main engine. When this is the case it is usually a centrifugal pump driven by a small engine. A **centrifugal pump** consists of a disc containing projections or vanes, so shaped that, when running in water, they impart such a velocity to the water that it is raised to some height. In Fig. 252 the shape of these vanes is shown at A, together with the direction of motion. As the water enters at the center B, flowing in from the sea, it is forced to rotate by the vanes as it moves toward the periphery of the disc. The result is to give it a high velocity. Since the distance between the vanes increases as they extend from the center, the height of the vanes from the disc is decreased toward the outside of the disc or **runner** to give a proper velocity relative to the disc. The quantity of water from the disc gradually increases as we approach the discharge C, because, at this point, the discharge from the whole circumference is passing, while at D there is only the small discharge from a portion of the runner; as points are taken further from D the quantity of water flowing in the passage E increases. The runner is made with vanes on each side, so that it is balanced, the water pressure on one side only causing an end thrust. The runner A is made of bronze and keyed to, or cast with, the bronze shaft F. This shaft usually runs in **lignum-vitæ** bearings, being lubricated by water. Bearings of this exceedingly hard wood are used in places where water can be constantly applied, as it forms a durable and efficient support under these conditions.

The engine, which is coupled to the flange G of the pump, is an ordinary D slide-valve engine without any governor, as the load is practically constant. The great value of such a

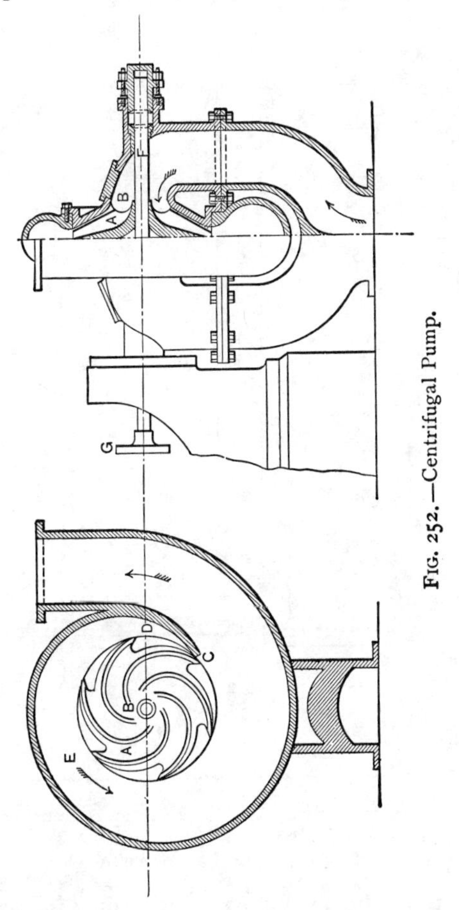

FIG. 252.—Centrifugal Pump.

pump is that it will throw large quantities of water under low heads, and do it with fair efficiency, using a simple runner which does not require the use of valves or any other check mechanism. As the water is taken below the water-line of the ship, and returned below that level, we do not raise the water, the only work done by the pump being that required to give the water the necessary velocity and to overcome the friction in the tubes.

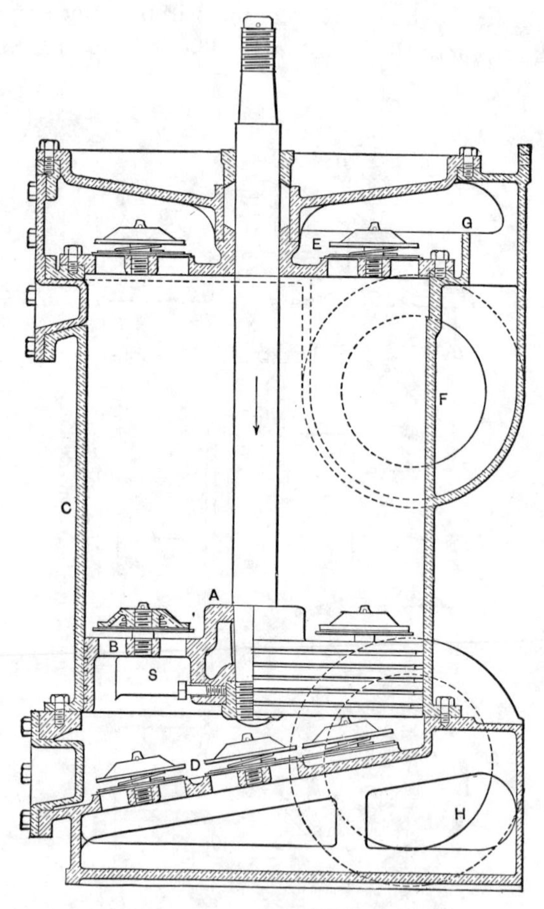

FIG. 253.—Vertical Air Pump.

Air Pump.—The air pump, which may be used with the surface condenser, consists of a brass **pump bucket** A, Fig. 253, which is in reality a piston pierced by the openings for the valves B, working within a brass cylinder C, at each end of which are sets of valves, the **foot valves** D and the **head valves** E. When the bucket is raised a vacuum is formed beneath, and the water, vapor, and air are drawn through the foot valves. These are placed on an incline to insure the removal of the vapor and air. The water and air, as the bucket descends, are prevented by these valves D from returning, and are driven through the valves B in the bucket. On again raising the bucket this mixture is driven out through the head valves, falling into the discharge pipe F after the water reaches the level of the lip G. By means of this lip the head valves are covered with water, and, since the other valves are so placed that they are covered, all the valves in this pump work drowned. It is found that this aids the action of the valves, giving a better vacuum. When there is not much air present small pet cocks are opened to prevent the blow which occurs from the water as it reaches the valves. To these cocks at each end of the cylinder are joined check valves opening inward, thus allowing a small amount of air to enter, which being gradually compressed raises the valves quietly. The pump suction pipe is connected at H, and the **discharge pipe** at F. The cylinder, or barrel, the bucket, the base and cap castings, and at times the valves, are made of brass, with the piston rod of rolled bronze. This is to avoid the corrosion which is apt to occur in such machinery when made of iron.

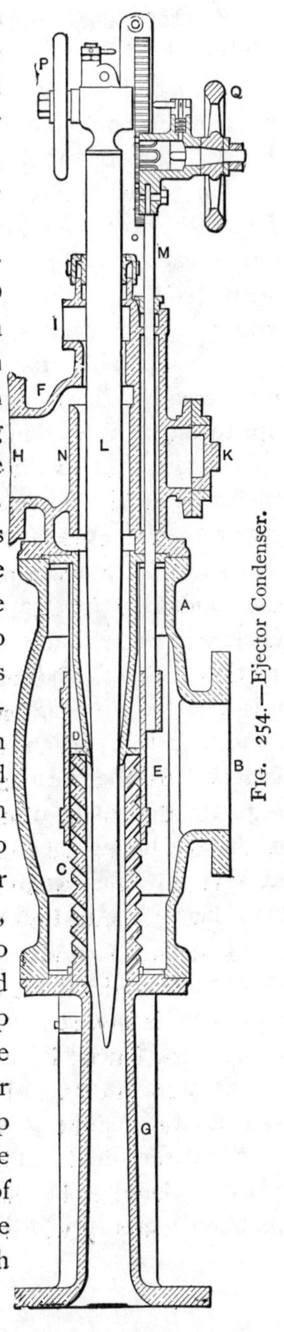

FIG. 254.—Ejector Condenser.

Ejector Condenser.—This form of condenser, as manufactured by L. Schutte & Co., is shown in Fig. 254. It consists of a body A, with an exhaust opening at B, in which are contained a combining tube C, a water nozzle D, and a sleeve E. On the top of the body is the water head F, and at the bottom is the tail piece G. In the head piece are openings H for the water suction, I for the starting jet, and K for the check valve. Through stuffing boxes the ram L and the sleeve rod M pass. To start the ejector condenser water under pressure, or steam, is allowed to enter through I, and this passing through the narrow passage around the ram creates a vacuum in the space N, drawing water up through a suction pipe, and delivering it through the space around the spindle. When this water discharges from the the water nozzle D it has a high velocity, and on coming in contact with the exhaust steam entering at B, the steam is condensed and a vacuum produced in the body A. After starting the condenser the steam or water at I may be shut off, and the energy in the exhaust steam is sufficient to maintain the water suction. Should the supply of water be too great or too small, it is adjusted by depressing or raising the ram by means of the hand wheel P, which turns a pinion which gears with a rack fastened to the head piece. The hand wheel Q serves to regulate the steam discharge. If the discharge is small, part of the combining tube is cut out. The check valve connected at K is only used when steam is employed to create the suction, being blanked off where water under pressure is used.

The ejector condenser maintains a vacuum without the use of buckets and valves, although at times, as with the injector, its action is not steady, the vacuum being lost and difficulty being experienced in restarting it.

Vacuum Gauge.—To measure the completeness of the vacuum the height of the mercury column which it will support is determined. Thus, if a glass tube connected to a condensing chamber be dipped beneath mercury, it will be found that the mercury will rise to a definite height. If the vacuum were perfect, this would be the barometric height of about 30 inches; but, since this is not possible, the height will be about 26 to 28 inches. Now since 2 inches of mercury corresponds to about 1 pound per square inch pressure, the absolute pressure of the steam with a 26-inch vacuum is $\frac{30-26}{2}$, or 2 pounds. To measure this pressure a Bourdon gauge, similar to that shown in Fig. 58, is usually employed, the scale being graduated to read in inches of mercury, the graduations extending from 0 to 30. Combination gauges are graduated to read above the atmosphere in pounds and below in inches, so that should an engine change from condensing to non-condensing the gauge would indicate the pressure.

Free-exhaust Valve.—As the vacuum in the condenser may be broken, or from some cause the condenser become clogged, it is necessary to have a safety device on the exhaust pipe between the engine and the condenser. Such a valve, called an automatic free-exhaust valve, is similar in action to the back-pressure valve shown in Fig. 109. From a tee in the exhaust line a branch extends to the atmosphere, and in this line the valve is placed, so that, should the condenser fail to act, it will open when the pressure is greater than that of the atmosphere.

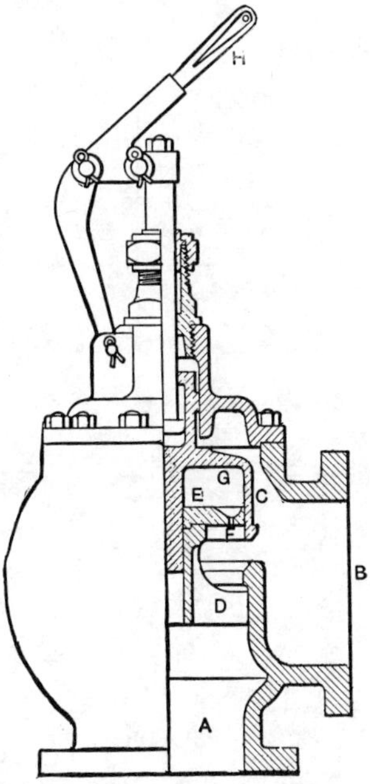

Fig. 255 is another type of automatic free-exhaust valve. The condenser connection is at A, and the free-exhaust pipe is at B. As shown in the figure the valve is open; the moving part of the valve being shown at C. The wings D, below the valve, support a plate E having small

FIG. 255.—Automatic Free-exhaust Valve.

holes F into the space under the valve. When the pressure in the condenser pipe is less than the atmosphere the valve C falls, the dash pot action of the space G causing it to seat quietly. The valve can be held open or closed by the external handle H.

Cooling tower.—Where a supply of water for condensing

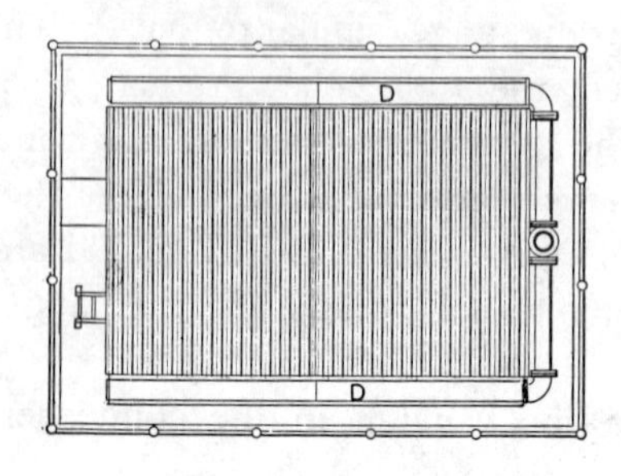

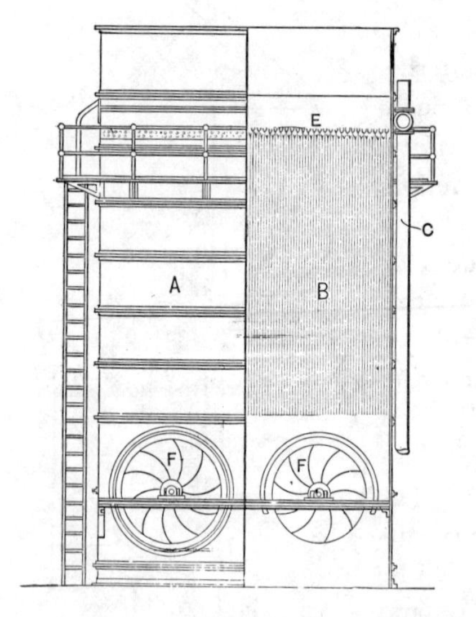

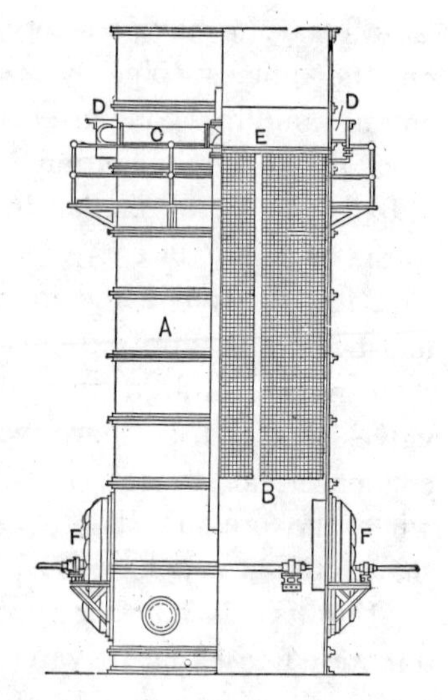

FIG. 256.—Cooling Tower.

purposes is not available most engines are run non condensing, the gain in efficiency not being equal to the cost of water. To run engines condensing under these circumstances, attempts have been made either to condense the steam by passing it through pipes, which have a large surface exposed to an air blast, or to cool the hot water from the condenser (jet, surface, or ejector) by allowing some of it to evaporate. This evaporation may be accomplished by discharging it in sprays from numerous jets, by allowing it to fall off a series of trays, by allowing it to circulate through a large pond, this depending on surface evaporation, or, lastly, by allowing it to fall in thin sheets through a current of air.

The first method has been used only for small engines, while the latter methods are employed for larger installations. Of these the cooling tower alone will be described. The cooling tower, Fig. 256, consists of a sheet-iron structure A, built of plates riveted to, and stiffened by, angle irons and containing a large number of wire screens B, called **mats.** The hot condensing or circulating water from the jet or the surface condenser is discharged from the pipe C into the troughs D, from which it discharges through small holes into the open pipes E, over the mats B. The water then drops to the bottom of the tower, where it is collected and passes to the condenser. The **fans** F, which are driven from an auxiliary engine, or electric motor, cause a strong current of air to pass up through these mats, cooling the water by the evaporation of part of it, and by heating the air. The evaporation of one pound of this water will abstract about as much heat as was acquired by the cooling water from one pound of the steam condensed, and, since the air which passes through the tower becomes heated, it also must abstract heat, thus reducing the amount of water to be evaporated. Hence it is seen that the water evaporated is less than the amount condensed, and, if the condensed steam from a surface condenser be used as boiler feed, the amount of water to be supplied from outside sources is less than the steam consumption of the engine. This steam consumption has also been reduced by condensing, so that the quantity of water used for driving an engine is greatly reduced. This enables a station to be run condensing with a smaller supply of water than a non condensing plant of the same output, even allowing for the power required to raise the cooling water to the top of the tower, which may be fifty feet, to run the fans and the air pump. In a test made on a power plant the power used with the main engine developing 2100 H.P. was 60 H.P. for lifting and circulating the cooling water, 8 H.P. for the air pump, and 41 H.P. for the fan. In this plant, condensing increased the power of the engine by more than 600 H.P. The power to drive the air and circulating pumps is very small compared with the great gain from the reduction of back pressure. For instance, with a marine engine developing 8500 H.P., the air pump will require about 69 H.P. and the circulating pump about 30 H.P.—the combined amount being a little over 1 per cent. of the total. The saving by the reduction of back pressure will amount to about 30 or 35 per cent.

Evaporators.—In many manufacturing processes, such as sugar-refining and ice-making, or at sea, it is necessary to evaporate water or other liquids, and in such cases evaporators are used. Fig. 257 shows the Lillie evaporator. It consists of a cast-iron casting A, with the head B on one end and the tube plate C and the head D at the other. The tubes E are 3″ copper tubes expanded into the tube sheet C, and having their outer ends closed except for small holes. The tubes F extend into the distributing box G, and have a small slot cut along their upper surface.

Steam entering at H passes into the tubes E, which are covered by water falling from the distributing tubes F. This water is pumped from the bottom of the float box K, by the centrifugal pump L, through the pipe M, to the distributing box G. The water abstracts heat from the steam on the inside of the tubes E, and the steam condensed from this cause falls to the bottom of the front head and escapes through a trap. The water or other fluid on the inside, however, as it extracts heat from the steam soon reaches the boiling-point and sends off vapor through the pipe N.

In distilling water for the manufacture of ice, this pipe is connected to a surface condenser and air pump, so that the boil-

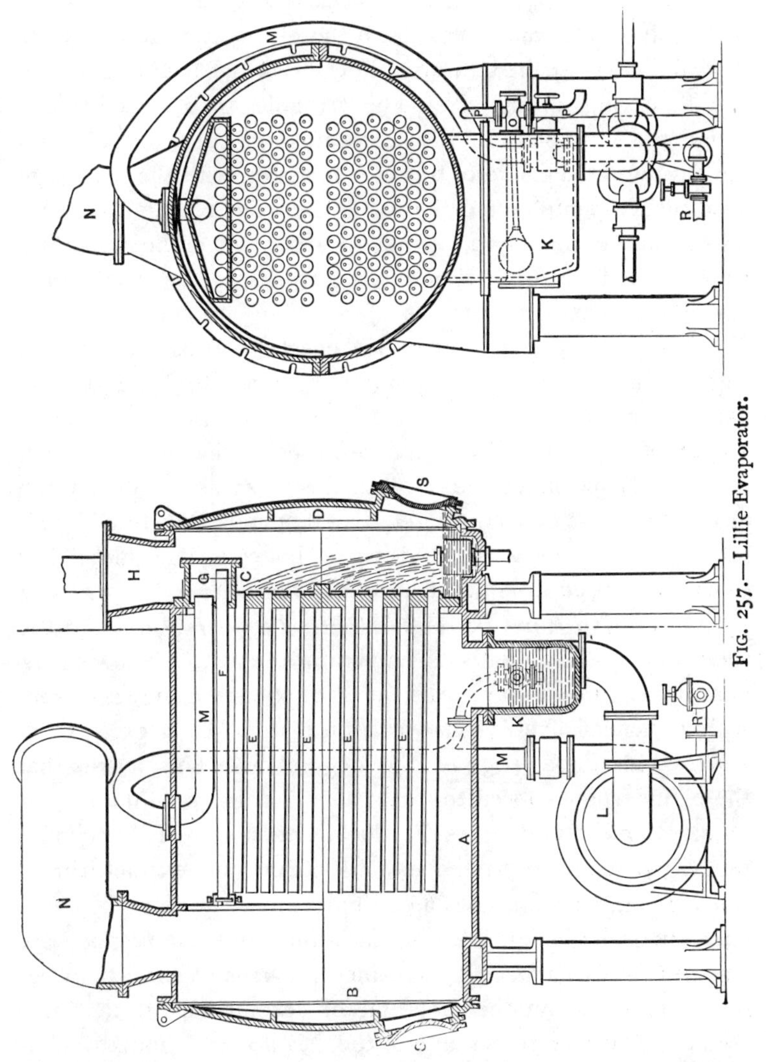

FIG. 257.—Lillie Evaporator.

ing-point of the water is low, and hence exhaust steam can be used on the inside of the tubes. The air in this steam is drawn by the air pump through the small holes at the end of the tubes. As the water boils away in the vessel A, the float in the box K opens a valve O, and allows more water to enter the pipe M through the supply-pipe P. When the liquid in the vessel is such that it becomes thick from the evaporation of the water the valve R is opened to discharge the heavy liquid, the fresh liquid entering through P. The manholes SS are used to examine the interior. In this apparatus the condensation from the steam on the inside of the tube is mixed with the discharge from the air-pump connected to N, and from 1 pound of steam we get approximately 2 pounds of water, since the amount of water boiled on the outside of the tubes is nearly equal to the amount condensed on the inside.

It is possible to increase this effect by using the exhaust vapors leaving through N in another evaporator, and the exhaust from this to a third. A single evaporator is called a **single effect,** while when three are used, as mentioned above, the name **triple effect** is used. In this way it is possible, when distilled water is desired, to get several pounds of distilled water by evaporating only 1 pound in the boiler. This method is extensively used in concentrating solutions.

Compounding and Multiple Expansion.—As the pressure of the steam falls its temperature decreases, and, in condensing engines using steam at 150 pounds gauge pressure with a vacuum of 26″, this variation is from 366 degrees F. to 126 degrees F., a range of 240 degrees F. This means that the cylinder is exposed to steam at 366 degrees F. during admission, and 126 degrees F. during exhaust. The cylinder walls give up heat to the exhaust steam and extract it from the incoming steam, causing initial condensation. To avoid excessive initial condensation, jacketing and compression are resorted to, as have been explained previously, and it now remains to show another method of accomplishing the same result. A larger percentage of the steam used is initially condensed in engines with early cut off than in those in which the steam has been admitted for a long period, say over one-half stroke. Moreover, in order to obtain the largest quantity of work from steam it must be expanded to a low pressure; otherwise a large portion of the energy which could be obtained by expansion is thrown away. To accomplish this expansion with high-pressure steam, expanding in a single cylinder, the cut off would be very early in the stroke, $\frac{1}{15}$ to $\frac{1}{20}$, and hence the percentage initial condensation would be very large. Now suppose steam at 165 pounds absolute be cut off at $\frac{1}{3}$ stroke and exhausted, after expansion, into a vessel, the pressure of which would be about 55 pounds. If from this vessel the steam be taken into another cylinder and cut off at $\frac{1}{3}$ stroke it would expand to $18\frac{1}{3}$ pounds, and if this were expanded in a third cylinder after cut-off at $\frac{1}{3}$ stroke, the final pressure would be 6 pounds absolute. Such an arrangement of cylinders form a **multiple expansion engine.** If the expansion is in two stages it is called a **compound engine,** in three stages a **triple-expansion engine,** and in four stages a **quadruple-expansion engine.** The names depend on the number of stages, and not on the number of cylinders used to attain the end. It is to be noted that, in the engine above, as the pistons are usually connected to cranks on the same shaft, and usually of the same throw, the area of the second piston, called the piston of the **intermediate cylinder,** must be three times that of the piston of the **high-pressure cylinder,** while the **low-pressure cylinder** contains a piston of an area nine times that of the first. In this case the range of the temperature in each cylinder is less than that which would occur were a single cylinder used; being 82 degrees F. in the first, 64 degrees F. in the second, and 90 degrees F. in the third. Other reasons for these engines being advantageous are to be found in structural considerations. The total expansion of the steam, or the **number of expansions,** is found by multiplying the several expansions in the different cylinders together, or it is the ratio of the volume of the low-pressure cylinder to the volume of the high-pressure cylinder at cut off. Were steam to act on the low-pressure piston and be cut off at the point corresponding to the reciprocal of the total number of expansions, there would be developed practically the same power as is actually developed by the engine. To withstand the full steam pressure in this large cylinder, however, thick cylinder walls, large piston-rods, crank, and wrist pins, and bearings would be required. By using a small diameter for the highest pressure we reduce the thickness of the metal for the cylinder and the sizes of all the parts, which is equally true for all the cylinders. Hence multiple expansion requires lighter parts. A most important advantage of these engines is that the **turning moment** on the shaft is more uniform, as are the strains set up in the engine frame. The pressure from the piston is transmitted to the connecting rod, being augmented by the guide pressure, and this force is divided at the crank pin into two forces, one acting tangentially to the crank pin, producing the tangential or turning moment, and the other in the line of the crank-arm, called the **radial pressure.** The latter of these produces only a pressure on the bearings, while the other causes the engine shaft to turn against external resistance. This tangential effort varies as the position of the crank changes, because of the relative directions of the connecting rod and crank, and also because the piston pressure varies, due to expansion. This causes the force received by the crank to change, so that a fly wheel is necessary to reduce the fluctuations. In most multiple-expansion engines the cranks are placed at such angles that, when one piston is exerting its maximum turning effort, the efforts from the others are small. In this way the turning effort is kept more nearly uniform, and in place of a great effort exerted at two points during a revolution, followed by periods of small effort, we have one in which there are several maxima which do not vary very much from the mean. This is very important where the engine is employed for certain operations.

To repeat, the reason for using multiple expansion are:

1st. The reduction in the size of the parts giving a lighter and less costly engine.

2nd. The production of a more uniform twisting moment,

giving smaller strains in the machine, necessitating a small fly wheel, and causing more uniform motion where a fly wheel is not employed.

3rd. Making the range of temperature less in each cylinder and probably reducing the initial condensation.

Compound Engines.—The first form of multiple expansion was accomplished in two stages in the compound engine. In this type of engine the steam may be distributed in either of two ways—one, in which the steam is exhausted directly from one cylinder into the other, first used in the Woolf engine, and the other in which the steam is discharged into a vessel called an intermediate **receiver.** The first is applicable when the pistons are connected to cranks which are 180 degrees apart, as in the Westinghouse compound engine described below, or when the pistons are attached to the same piston rod, as in the Ball engine, shown in Fig. 262. When the cylinders are arranged side by side with the cranks at 90 degrees, giving what is known as the **cross compound** engine, a receiver must be used. Cylinders in line form a **tandem compound** engine.

Westinghouse Compound Engine.—The construction of this engine is seen in Fig. 258. On a cast base A are bolted the bearings BB, the cylinder casting C, and the sleeve D. On the top flanges of the cylinders, which are cast together, is bolted the valve casing. The cylinders are cylindrical castings with flanges only, the steam passages being cast in the steam chest. The pistons EF are of the trunk type, being made deep to have ample bearing surface. They contain the wrist pins GG. The piston valve H is driven from the eccentric I by means of the bell-crank lever K. The eccentric rod is attached to this lever by the ball-and-socket joint M, because the axes of rotation of the lever and rod are at right angles. The eccentric I is turned spherical to permit the side motion due to the movement of the bell-crank lever. The piston valve H, which, like the pistons, is packed with spring rings, moves within the liner N containing the various ports. O is the steam port, allowing steam to enter from the steam space P.

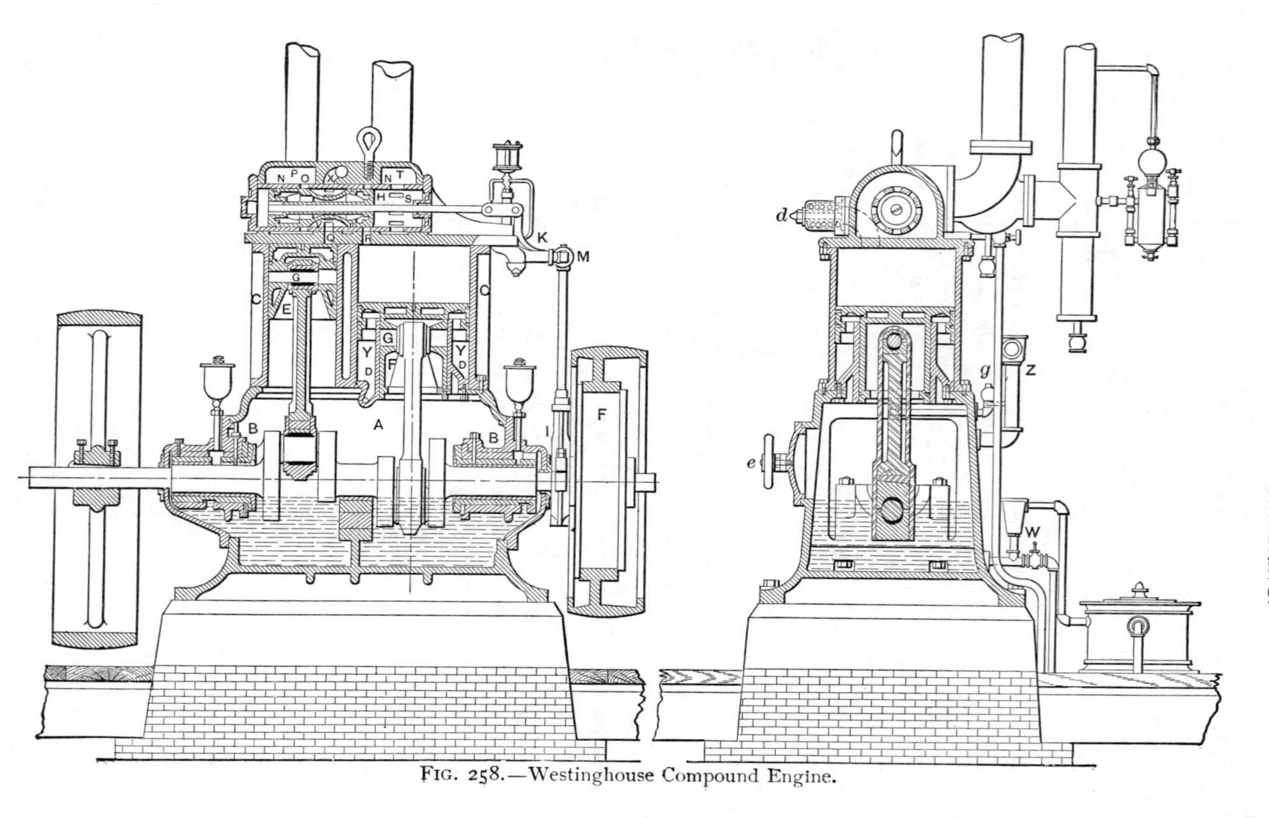

FIG. 258.—Westinghouse Compound Engine.

Q is the high-pressure cylinder port, R the low-pressure cylinder port, and S the exhaust port leading to the exhaust chamber T.

If the high-pressure piston is at the top of the stroke, the low-pressure piston will be at the bottom of its stroke, the cranks being 180 degrees apart. The valve H, having lead, allows steam to enter through O filling the clearance space V around the valve and in the top of the cylinder, and forces the piston downward. The port R being uncovered, steam exhausts from the low-pressure cylinder. As the engine turns, the valve moves to the extreme left of its travel and then begins to return, finally covering O, and, a little later, R is closed. The first cuts off the steam from the high-pressure cylinder, allowing the steam in the cylinder and the space V to expand, while the second action causes the steam in the low-pressure cylinder to compress. When the piston E reaches the lowest point in its travel the valve has moved to the right so far that Q and R are in communication. This allows the steam from the small cylinder to enter the large cylinder, and as its pressure is exerted here on a larger piston, the large piston moves downward, driving the small piston upward. After the valve moves to the right it again returns to the position where Q is cut off from R, causing the low-pressure steam to expand alone, while the steam in the high-pressure cylinder is compressed until live steam is again admitted. The opening X is a **by-pass connection,** admitting steam from the steam chest P into the space V, thus admitting high-pressure steam to the low-pressure cylinder for starting.

The base A is partly filled with oil and water, so that, as the cranks turn, the splashing of this oil lubricates the various parts. If the level of this mixture rises, from steam which condenses after leaking into the case, the water escapes from an overflow W, which is attached to the lowest part of the case but extends upward to the water level before discharging. Any steam that is not condensed in the case is discharged through the vent Z. It will be noticed that the piston F has an additional portion which moves in the sleeve D. This is placed in the cylinder to prevent the oil and water from being driven out of the case. As the two pistons are of different diameter the volume in the base, beneath the pistons, would be continually changing if this sleeve were not used, being a maximum when the high-pressure piston is at its lowest point. The continual change would tend to blow the oil from the overflow W, and air from the vent Z, but, if this sleeve is made of the same diameter as the high-pressure piston, the volume remains practically constant. The space Y is called the cushion space, as the air within this is compressed and expanded during each revolution. The valve G serves to drain the space Y, or to allow the air to escape when turning the engine by hand.

As the engine is single acting the pressure on the pistons is always downward, the connecting rods are always in compression, and the pressure on the bearings is upon the lower block. The connecting rod, Fig. 259, is made of a central rod B and the strap A, with boxes at each end. The rod carries all of the load, the strap taking only the initial stress put in by the wedge D to keep the rod tight. The bolts HH hold the lower box to the strap A, and the bolt K draws the wedge D so as to force the boxes C and F against the pins. The rod B is held in place only by the flanges of C and F, which also extend over the strap A. The bearings of this engine, Fig. 260, which are bolted to the frame of the engine, are made of a cast-iron shell A and flange B, within which are placed the brasses CD. The upper brass D is held in place by the tap bolts EE, while the wedges FF are used to take up the wear below. The oil from the cup G is distributed over the journal, and finally falls into the bottom of the shell and then to the base of the engine. The small sleeve or collar H is used to prevent the oil from work-

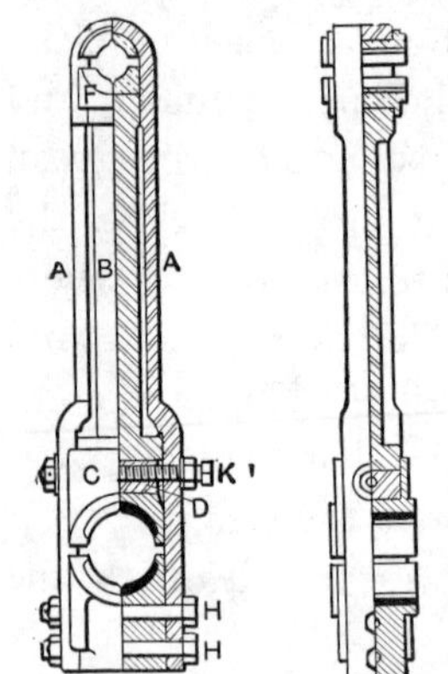

FIG. 259.—Connecting Rod, Westinghouse Engine.

ing out of the bearing. Any oil that creeps out on the shaft is thrown off by centrifugal force when it reaches this collar, and then falls to the bottom of the shell. The center bearing

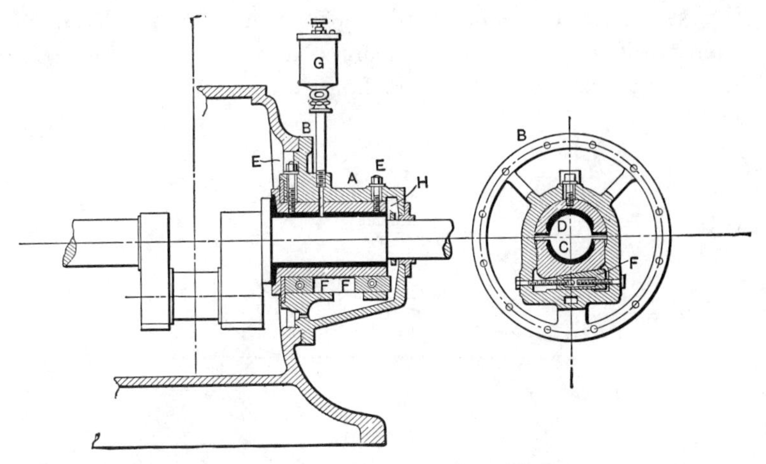

FIG. 260.—Bearing, Westinghouse Engine.

is adjusted by a wedge, the top brass being held beneath a yoke. The crank shaft with the inside cranks is made of one forging of the best steel.

There is attached to each cylinder a relief valve *d*, Fig. 258, which relieves the cylinders of water. Plugged holes in the end of the valves are provided for the attachment of indicators when testing. The connecting rods and bearings can be adjusted through the doors *e*, Fig. 258. The governor controlling the eccentric is inclosed in the box F in the fly wheel. The lubricators used on the engine are four in number in addition to the oil bath in the base, one large oil cup over each bearing, one cup to lubricate the rocker and eccentric, and a cylinder lubricator on the steam pipe. The eccentric receives its oil through a passage leading from the axis of the bell crank to the ball, and then through the hollow eccentric rod. A guard, not shown in the figure, catches the drip from the eccentric. The lagging around the cylinder in this figure is made of sheet iron.

This engine has given good satisfaction, costing little for repairs, and having a low steam consumption. The diagrams taken from each cylinder of one of these engines are shown in Fig. 261 upon the same card. The figure shows diagrams taken at different loads, the steam consumption varying from 22.4 to 29.2 pounds per I.H.P. hour.

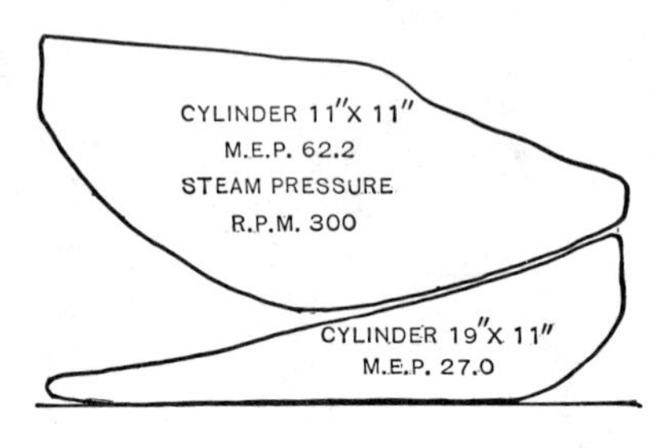

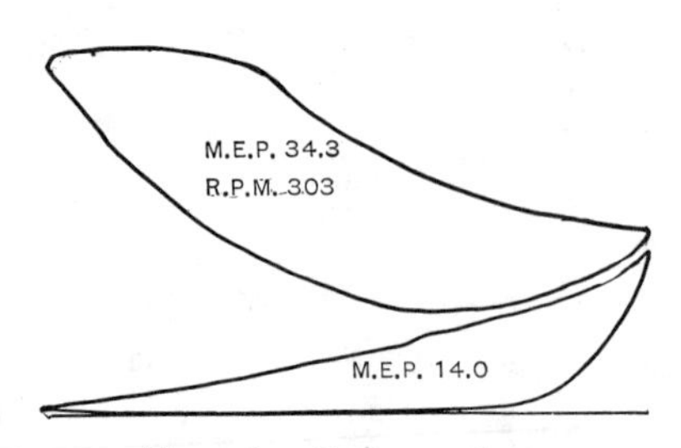

FIG. 261.—Indicator Cards.

Tandem Compound Engine.—Fig. 262 shows the construction of a tandem compound engine, made by the Ball Engine Co.

The cylinders A and B of such engines are arranged with their axes in one straight line with a distance piece C between them, the object of which is to have the stuffing boxes accessible. In many engines this is made separate from the heads, and not as shown in the figure. The two pistons D and E are attached to a single rod F, which is reduced in section behind the low-pressure piston. In the engine here shown the steam chest of the high-pressure cylinder is on the front side in the figure, while

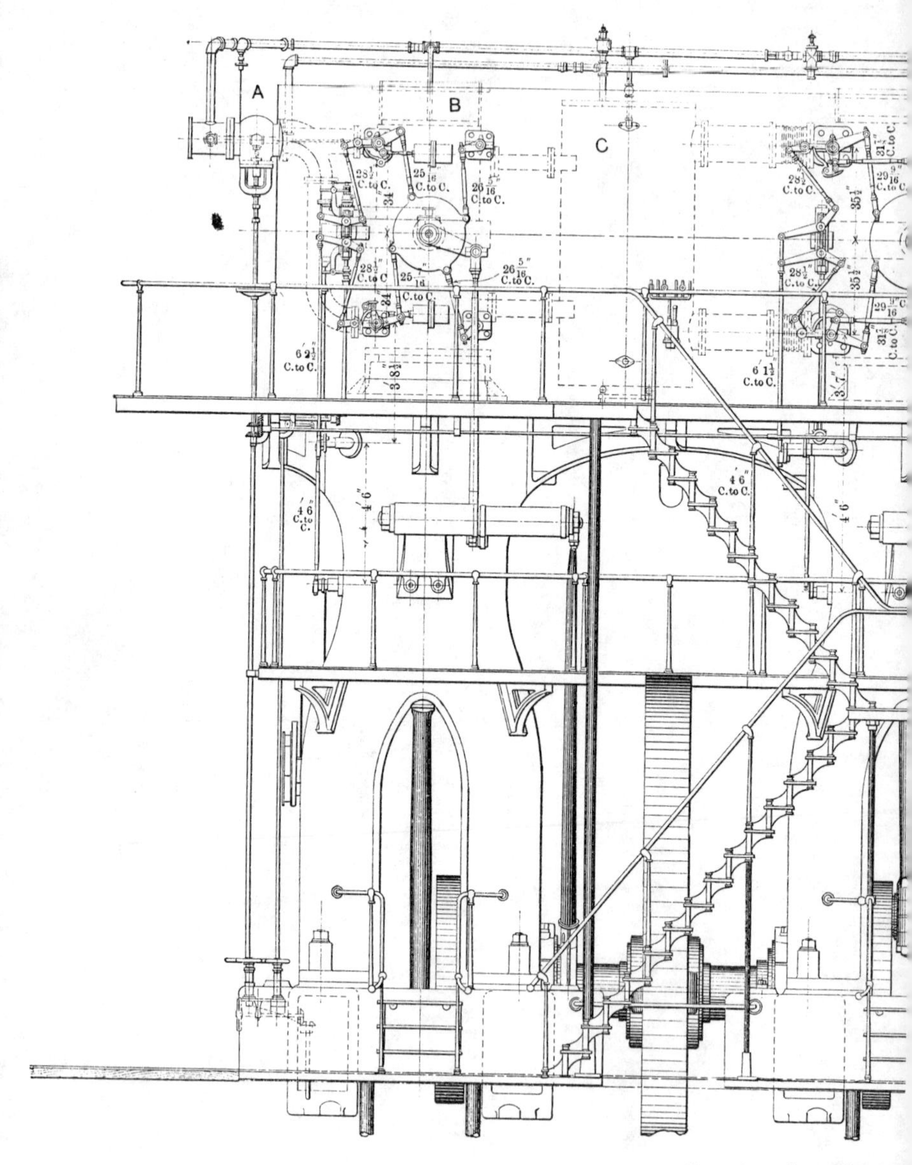

FIG. 263.—Snow Triple-

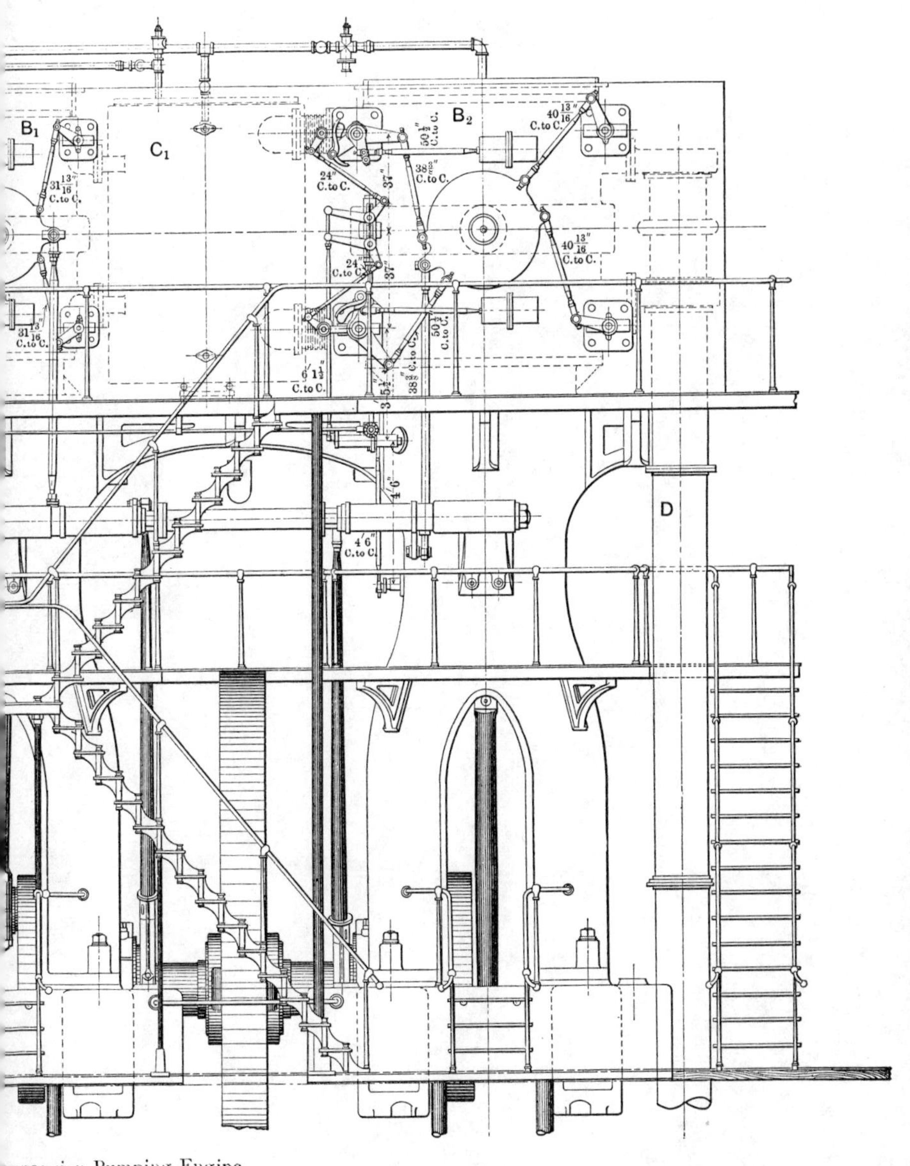

expansion Pumping Engine.

(*To face page* 255.)

the low-pressure cylinder carries its steam chest on the opposite side. The two valves are worked by separate eccentrics, the high-pressure valve being under the control of the governor. The exhaust from the high-pressure cylinder passes into the low-pressure chest through the pipe G. In many cases the valves of a tandem engine are attached to the same valve rod.

The small cocks HH are the drain cocks, and the holes on the upper side capped by the plugs KK are for the attachment of the indicators.

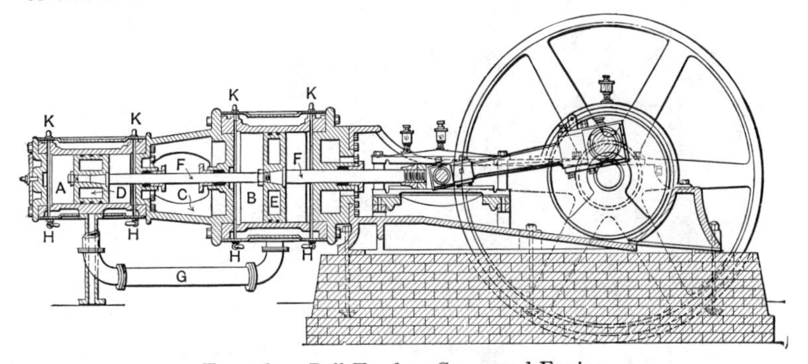

FIG. 262.—Ball Tandem Compound Engine.

Triple-expansion Engine.—A triple-expansion engine of the receiver type, used for pumping water, is shown in Fig. 263, its action being similar to that of a compound-receiver engine. The engine shown is that built by the Snow Steam Pump Works * for the Indianapolis Water Co. One view only of the engine is shown, as it clearly indicates the path of the steam. In it the steam is taken through the main throttle valve A to the high-pressure cylinder B, the admission being controlled to all cylinders by valves of the Corliss type. The steam is exhausted into the receiver C, where it is warmed by steam passing from the boiler through a large coil. The receiver C, or, as it is called in this case, the **reheater,** dries the exhaust steam and superheats it before it enters the intermedi-

---

* Transactions A. S. M. E., Vol. xxi.

ate cylinder B_1. The **second reheater** C_1, which contains a larger coil of pipes filled with high-pressure steam, serves to dry and superheat the steam entering the low-pressure cylinder B_2. These reheaters are connected to one cylinder by nipples and flanges, while the connections to the second cylinder are by means of a corrugated pipe, to allow for expansion. The exhaust of the low-pressure cylinder is conducted through the pipe D to a surface condenser, through which the circulating water is drawn by the main pump suction. The cranks of this engine are set at 120 degrees apart, to make the turning effort uniform, the middle crank being an inside one. The cranks being at this angle necessitates the use of a receiver between the cylinders, but this receiver is not usually so large as the reheaters shown. In many engines the pipes connecting the cylinders act as receivers. The barrels and heads of the cylinder are jacketed to reduce the condensation. While jackets and reheaters are fitted on most large engines, it is not always certain that their use reduces the steam consumption. Experiment shows that reheaters are of little value if they do not superheat the steam.

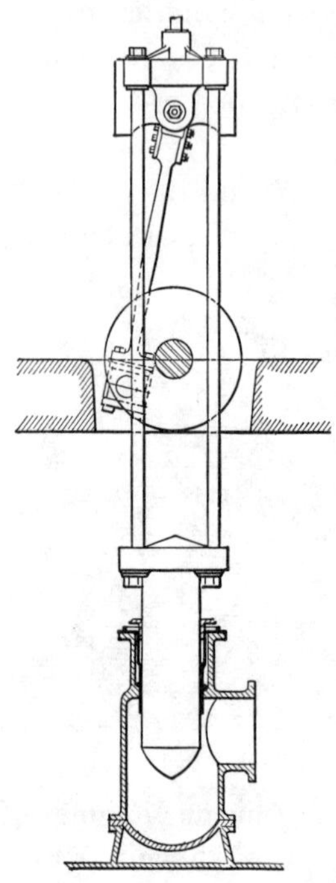

FIG. 264.—Pump Connections, Snow Engine.

The cross heads of these engines are large square blocks, to which, beside the piston rod and connecting rod, four pump rods are attached, which connect with the pump plunger and are used to span the center crank, as shown in Fig. 264. This figure shows the pump cylinder and the single-acting plunger, together with the cross heads. The method of packing is clearly illustrated. The valves are not shown in this figure, as they were placed in a separate vessel connected to the pump cylinder by a large pipe.

The fly wheels are between the cylinder frames, which are modified A-frames so common on marine engines. Most pumping engines are vertical, as this not only saves floor space but reduces the friction load. The weight of the parts, requiring much of the energy developed by the piston on the up stroke, acts in conjunction with the steam pressure on the down or pumping stroke in forcing the water. Traps are provided to drain the condensed steam from the jackets and reheater coils.

The cards from this engine are shown in Fig. 265, and were taken by Prof. W. F. M. Goss during a test. The steam

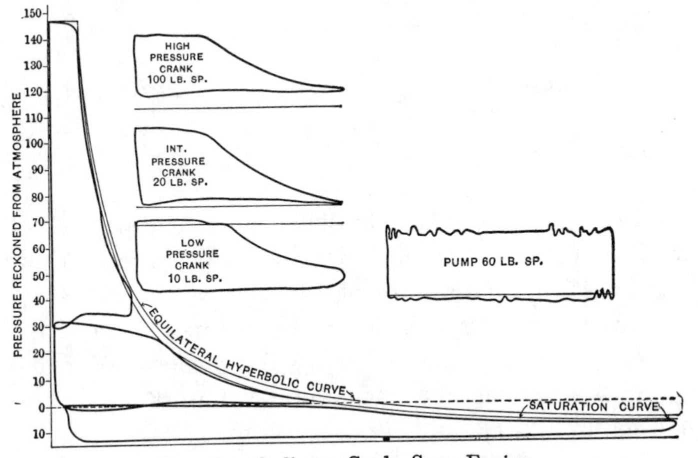

FIG. 265.—Indicator Cards, Snow Engine.

pressure at the throttle valve was 154.6 pounds by gauge, in the first receiver it was 32.3 pounds, and in the second receiver 1.7 pounds. The vacuum was 12.7 pounds, being 26″ of mercury. These cards show the effect of discharging into the receiver, as the exhaust line rises on account of the compression of the steam, and then on admitting steam into the next cylinder the pressure falls. It will be also noted that the steam in the high and intermediate cylinders expands almost to the pressure of the receivers, there being practically no **free expansion.** This free expansion occurs whenever steam is allowed to discharge from a space in which the pressure is high into one in which the pressure is much lower. These cards are drawn to different spring and volume scales, and to reduce them to the same scale, as well as to compare the net work from these cards from that which might occur in a single cylinder, the **combined card** is also drawn. In this combined card the clearance lines of each card are placed in line, and the length of the cards, which represent volume, are drawn to the same scale. The rectangular hyperbola and a curve for saturated steam is drawn, starting with a point just after cut off. This saturation curve shows the condensation and re-evaporation in the various cylinders. The indicated horse power from the different cylinders were as follows:

High-pressure .... 257.8 I.H.P.
Intermediate...................... 224.5 I.H.P.
Low-pressure...................... 300.8 I.H.P.

A typical card from one of the pumps is also shown. This card is similar to most cards obtained on pump tests, showing the fluctuations of pressure at the end of the stroke. The power from the water cylinders amounted to 736.3 I.H.P.

Duty.—The manner of stating the efficiency of pumping engines is by giving the amount of useful work in foot pounds done for each 1,000,000 British thermal units used. This is called the duty of the pump, and in the engine just described amounted to 147,500,000. That is, 147,500,000 foot pounds of useful work were obtained from 1,000,000 British thermal units. It will be remembered that the British thermal unit is the heat required to raise the temperature of one pound of water from 62 to 63 degrees F., and is the equivalent of 778 foot-pounds of work. The steam consumed amounted to 11.38 pounds per I.H.P. hour, and of this amount 1.3 per cent. was used in the reheaters and 4.1 per cent. in the jackets. The

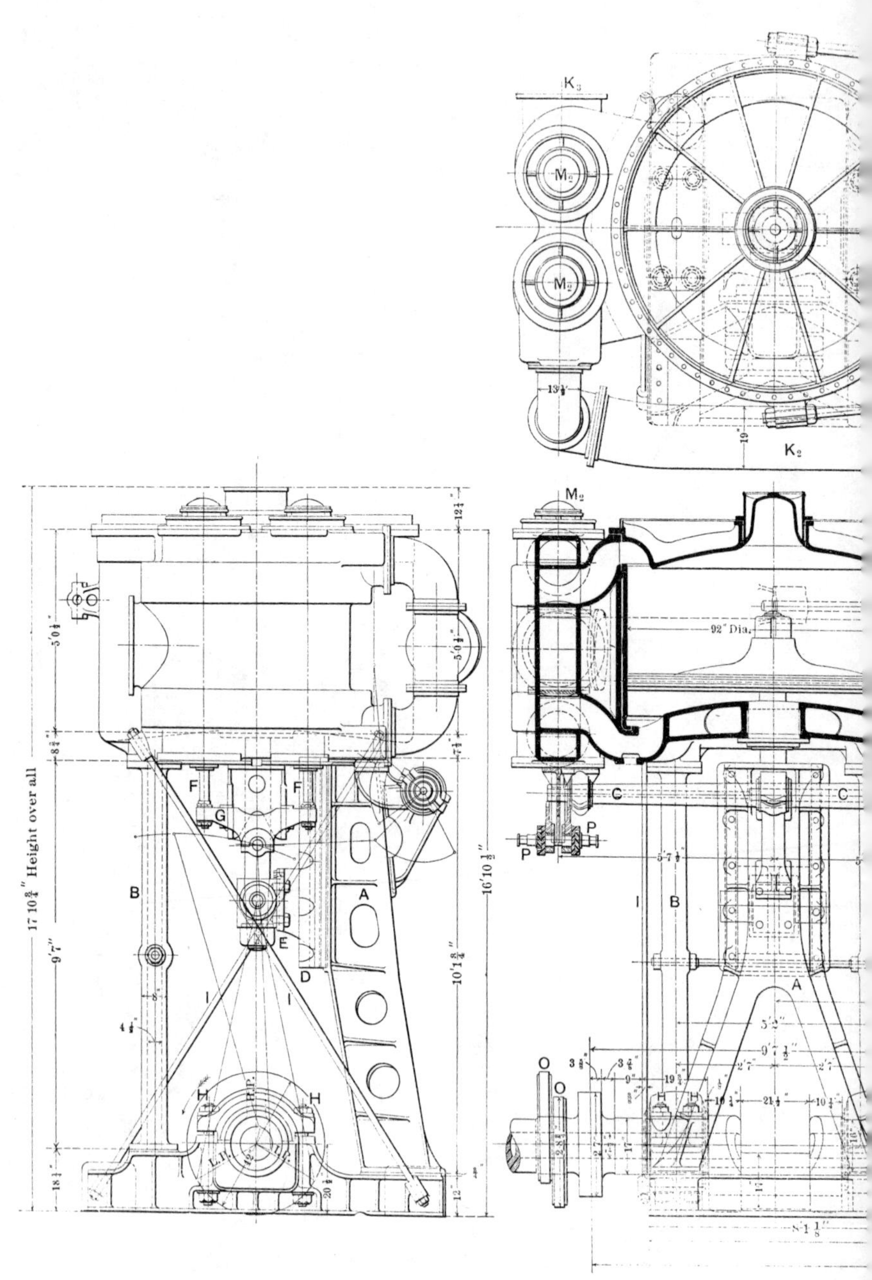

K₃
M₂
M₂
13¼"
19"
K₂
M₂
17 10¾" Height over all
5'0¼"
12¼"
8¾"
7¼"
16'10½"
10'1⅝"
9'7"
18¼"
F
F
G
B
A
E
D
I
I
H
H
92" Dia.
C
C
P
P
5'7½"
B
A
5'2"
9'7½"
O
O
2'7"
2'7"
19¾"
21½"
10¼"
17"
8'1⅛"

FIG. 2

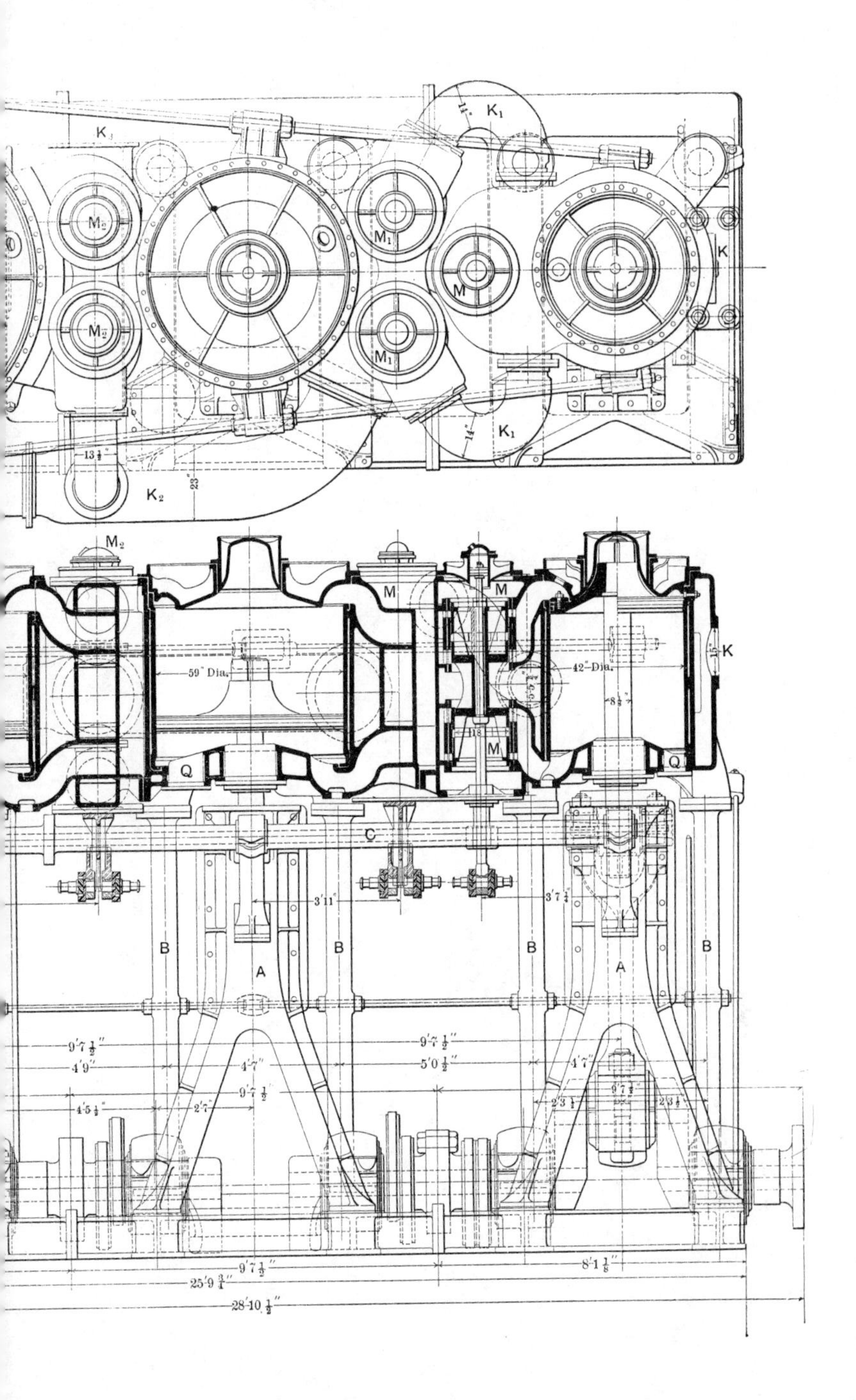

K3
K1
K
K1
K2
M2
M1
M
M2
M1
13 1/2"
23"
14"
M2
M
M
M
K
59" Dia.
42" Dia.
Q
Q
C
3'11"
3'7 1/4"
B
B
B
B
A
A
9'7 1/2"
4'9"
4'7"
9'7 1/2"
5'0 1/2"
4'7"
9'7 1/2"
9'7 1/2"
4'5 1/2"
2'7"
2'3 1/4"
2'3 1/4"
9'7 1/2"
8'1 1/8"
25'9 3/4"
28'10 1/2"

capacity of the pump was 20.3 million gallons in 24 hours, the builders' rating being 20 million gallons. This method of giving **capacity in gallons per twenty-four hours** is the common method of rating pumping engines.

Marine Triple-expansion Engines.—Fig. 266 shows the section through the cylinders of an engine for a cruiser of the U. S. Navy.* The cylinders are carried on A-frames A at the back, and on the steel columns B in front. The section shown is taken through the cylinders only, the remainder of the figure showing the back of the A-frames. The reverse shaft C controls the position of the Stephenson links, as the hangers are attached to the arms extending from this shaft. The side elevation shows the guides D and cross head E, as well as the two valve rods FF, which join to a cross head G that is attached to the block in the Stephenson links. The main bearings are shown in the side and front views. Four bolts, HH, are used to hold the caps in place. The bottom box is rectangular and fits into a similar receptacle in the bed plate of the engine. The **tie rods** I serve to stiffen the framing. Steam enters the high-pressure cylinder at K from the boilers, passing into the outside of the single valve chest M. From the center of this valve it enters the second cylinder through the bent pipes K_1K_1, and the two valve chests M_1M_1. From the center of these valves it is then conducted to four valve chests M_2M_2 by the pipe K_2. The steam then exhausts through the two openings K_3K_3. The pipes and valve chests act as receivers.

The method of forming the crank shaft is clearly seen in the figure. The sections are made interchangeable, so that a single spare section may be carried to replace any one, should it break. The dotted lines in the shaft and pin show the hole extending through the center. The eccentrics OO are shown in position on the shafts, and directly above them are the pins PP on the links to which the eccentric rods are attached. The

---

* Report of Engineer in Chief Geo. W. Melville, U. S. Navy.

pins shown on the rock-shaft end of the link are made long so that the hanger rods may be attached to the outer end of them.

It will be noted that the heads of the cylinders conform to the outline of the pistons, and that to the central part of the cylinder-head caps are bolted, forming manholes, by which the interior of the cylinder may be examined. Openings Q in the lower head are also made for the same purpose. The stuffing boxes of this engine are made separate from the cylinder, the castings being bolted into the head after a close fit has been made on two projecting rings. The valve shown is double ported on the steam side, but single for the exhaust.

The arrangement of the cylinders of marine engines is not always similar to that shown. For instance, the engines of the S.S. Campania are of the triple-expansion three-crank type, with two high-pressure, two low-pressure, and one intermediate cylinder, a high-pressure and a low-pressure cylinder being in tandem. This arrangement, shown in Fig. 267, permits each crank to do the same amount of work. In other engines there will be two low-pressure cylinders acting on separate cranks. In this case the exhaust from the intermediate cylinder is delivered to both of these. This gives us a four-crank triple-expansion engine.

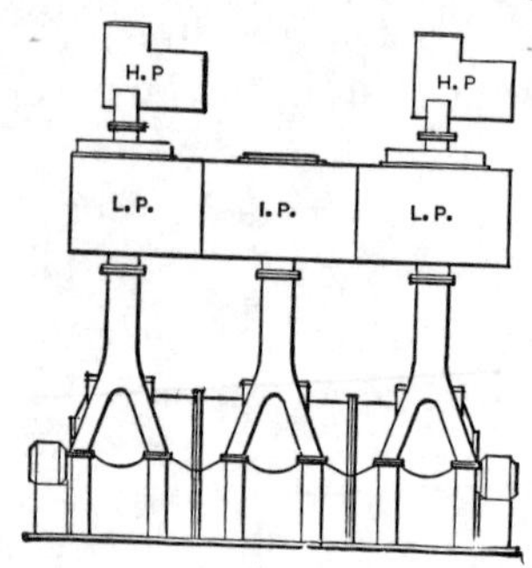

FIG. 267.—Cylinder Arrangement, Campania.

Quadruple-expansion Engines.—When the expansion of steam takes place in four stages we have a quadruple expansion engine. Fig. 268 shows the arrangement on the cylinders of the S.S. St. Louis and St. Paul. The high-pressure cylinders, which are in tandem with two low-pressure cylinders, exhaust into the first intermediate cylinder, and this in turn into the second intermediate. From this point the steam is exhausted into the two low-pressure cylinders.

The motion of the reciprocating parts of an engine cause vibrations which are quite apparent on vessels. To reduce this vibration the arrangement of cylinders and cranks is such that the effects are partially neutralized and the vibration reduced to a minimum. This has given rise to the various arrange-

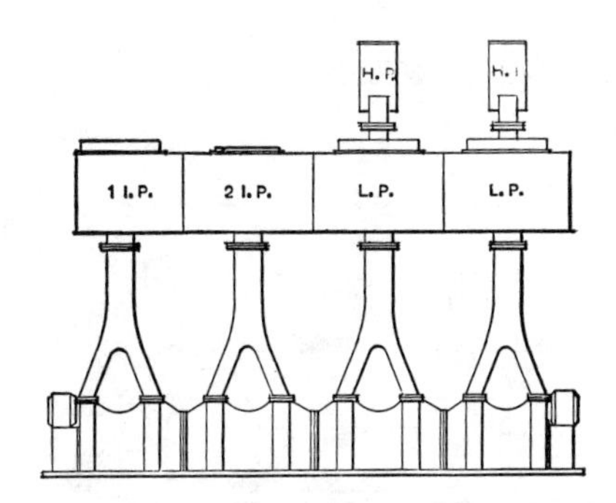

FIG. 268.—Quadruple Expansion, S.S. St. Louis.

ments of cylinders, and the order in which they act on the cranks and the angles between them.

Compound Locomotive.—The compound locomotive has been made in many forms. Sometimes the steam is admitted to the cylinder on one side of the engine and exhausted into the one on the opposite side, the latter being of larger diameter. In some English engines both outside cylinders exhaust into a third cylinder, which is placed between these and acts on an inside crank on the front axle. The arrangement of the pistons and valve of a Vauclain compound locomotive is shown diagrammatically in Fig. 269. This system is designed to obtain the economy due to compounding, and at the same time develop equal powers on the two sides of the locomotive, using practically the same methods of handling that are in use with single-expansion engines. A high- and a low-pressure cylinder in a single casting are used on each side of the locomotive, the pistons of which are connected to a common cross head, while a single piston valve controls the events for both cylinders. The action of the steam in this system is as follows: Steam is admitted outside of the piston valve A, and, when the valve is moved to the right, enters the left end of the high-pressure cylinder. This action allows steam to exhaust from the right-hand end of the high-pressure

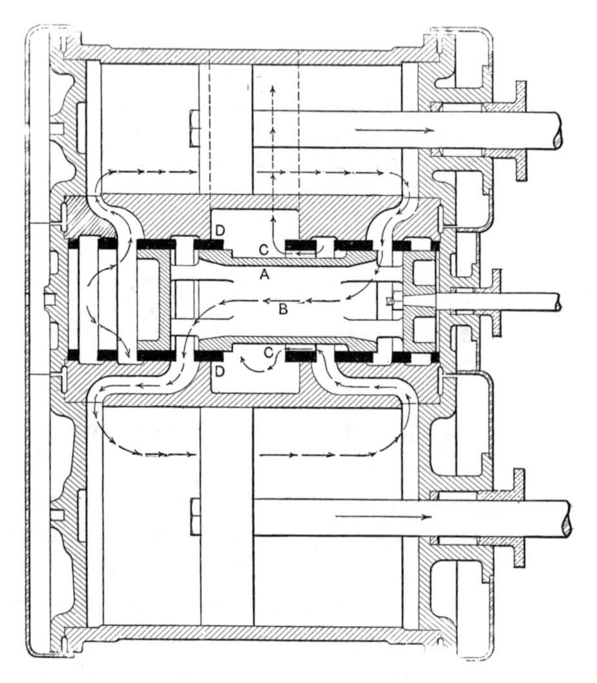

FIG. 269.—Vauclain Compound Locomotive.

cylinder through the hollow space B, in the center of the valve A, to the left-hand side of the low-pressure piston, while steam on the right escapes through the exhaust cavity C around the valve. At the proper time steam is cut off from the high-pressure cylinder and expansion takes place. This is followed by the closing of the exhaust on the other end of the high-pressure cylinder, which cuts off the steam in the left end of the low-pressure cylinder, and hence expansion occurs here also, while compression takes place in the right-hand end of the high-pressure cylinder. The closing of the exhaust from the right-hand end of the low-pressure cylinder causes compression. The further movement of the valve admits steam to the other side. After boring the cavity for the valve a liner D is forced in. This is represented by the heavy black lines around the valve. In addition to the usual drain cocks and relief valves found on any cylinder, two air valves are placed in the steam passage leading to the low-pressure cylinder, and a starting valve is furnished connecting the two ends of the high-pressure cylinder. The object of the first of these is to allow air to enter the low-pressure cylinder should expansion take place below the atmospheric pressure. This is necessary, because the vacuum produced would draw cinders into the cylinder through the blast pipe when the exhaust port opens the cylinder to the atmosphere. The starting valve is for the purpose of admitting live steam to the low-pressure cylinder on starting. It allows steam to pass from that side of the high-pressure piston on which admission occurs to the opposite side, which is in connection, through the valve, with the low-pressure cylinder. In this way both pistons receive steam. The high-pressure cylinder is not very effective, as the pressure on each side is almost the same, but the large low-pressure piston, which is usually about three times the area of the high-pressure piston, receives a high pressure, and thus exerts a large starting force. The steam distribution in this double-acting engine is similar to that of the Westinghouse engine.

Steam Turbine.—One of the latest applications of the use of steam for the developing of mechanical energy is found in the De Laval and the Parsons steam turbine.

This method of using steam was suggested several centuries ago, but it has been only within the last twenty years that the idea has been developed so that it is of commercial value. The De Laval turbine uses the reaction caused by a jet of steam of high velocity impinging on a single set of radial vanes, while in the Parsons turbine the steam at a lower velocity acts on a series of stationary and moving vanes.

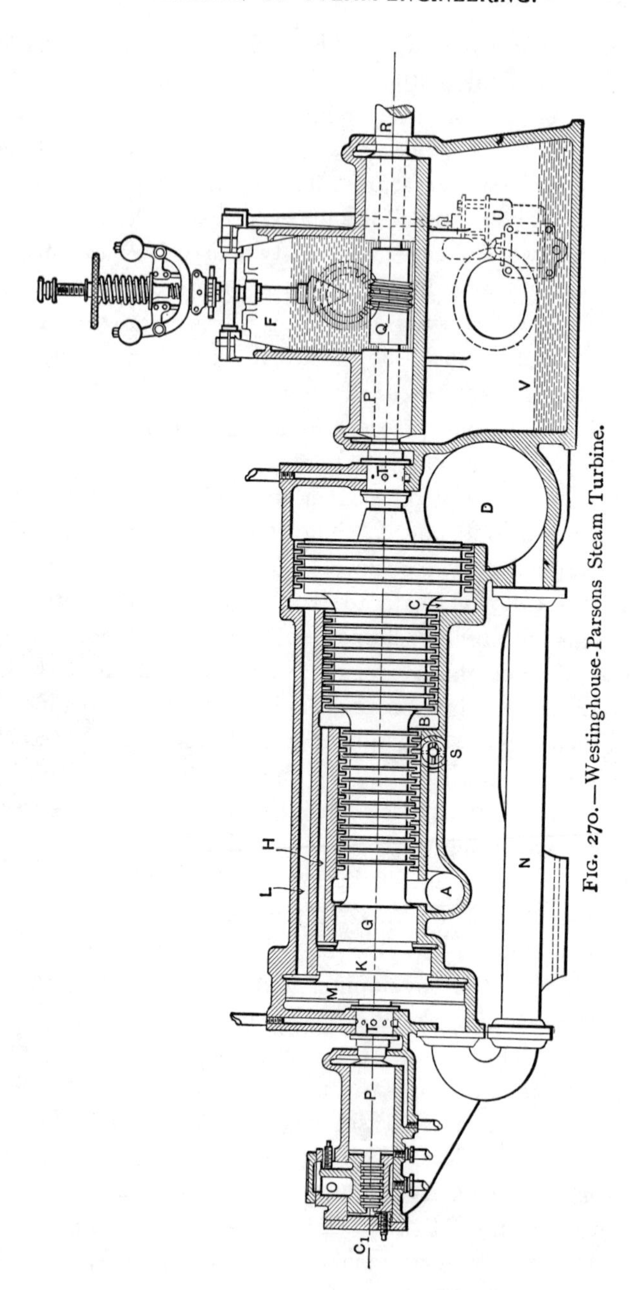

FIG. 270.—Westinghouse-Parsons Steam Turbine.

Fig. 270 shows a section through a large Westinghouse-**Parsons steam turbine.** Steam enters at A and passes among a series of fixed and movable vanes to B, where it enters another series extending to C. At this point begins a third set, the steam finally leaving at D, from which it is conducted to a condenser. The part F contains the governor, which gives an intermittent supply of steam. The admission is controlled by an eccentric, which, through a system of levers, raises and lowers a relay valve, which opens and closes the main admission valve. The governor acts on one of the levers in the system controlling the relay valve, and alters its motion so that it is raised a greater or less amount, and hence keeps the main valve open a corresponding length of time. This intermittent motion prevents the mechanism from sticking and renders its motion positive.

After entering the space above A, Fig. 270, the steam is directed by a set of fixed blades *a*, Fig. 271, against a set of movable blades *b*, and, on being discharged from these, it is directed by a second set of fixed blades *c*, against the next set of movable blades *d*. This steam acts in this way on twelve sets of blades until the space B is reached, and at this point the

FIG. 271.—Guide and Moving Blades.

diameter of the vane barrel is increased to allow the steam to move at a higher speed.

The steam acting on the blades causes the barrel between A and B to be forced to the right. There is a balance piston G on which the pressure at A can act.

The pressure at B is conducted by the balance port H to the balance piston K, to care for the end thrust from the intermediate blades, and by means of L and M the thrust from the low-pressure blades is cared for. These pistons are smaller by a small amount than the cylinders in which they turn, so that there is no friction, and yet the amount of clearance is such that the leakage is small. The pipe N connects the back of the piston M with the exhaust chamber D. The blades of the turbine are calked in grooves made in the barrel. In a 300 K. W. turbine the number of these blades is 31,073, of which 16,095 are moving blades.

The by pass S admits high-pressure steam to the intermediate vanes increasing the power. At O there is a thrust bearing which serves to keep the parts in proper position. The main bearings PP are made of four gun-metal sleeves, so constructed that the shaft may be self-centering. Q is the coupling connecting the shaft of the turbine to the shaft R of the machine to be driven.

To prevent air from reaching the space D, steam is allowed to reach the packing glands T at a pressure a little above the atmosphere. This, although it leaks in, prevents the air from entering and so assists in maintaining the vacuum. Oil is also introduced into these glands.

The oil for the journals is pumped by U into the chamber F, and reaches the reservoir V after passing through the bearings.

The steam consumption of a 300 K. W. Westinghouse-Parsons turbine is quite low, being 16.4 pounds of steam per hour per electrical horse power delivered from the dynamo. This would give about 14 pounds of steam per hour per I. H. P. During the test at the Westinghouse shops the data was as follows: Speed 3600 R. P. M., steam pressure 125 pounds, vacuum 26″ to 27″. These results indicate a great future for this form of motor.

The arrangement of the **De Laval Steam Turbine** with a dynamo is shown in Figs. 272 and 273. Steam in the steam chamber A issues from a nozzle B upon the vanes of the wheel

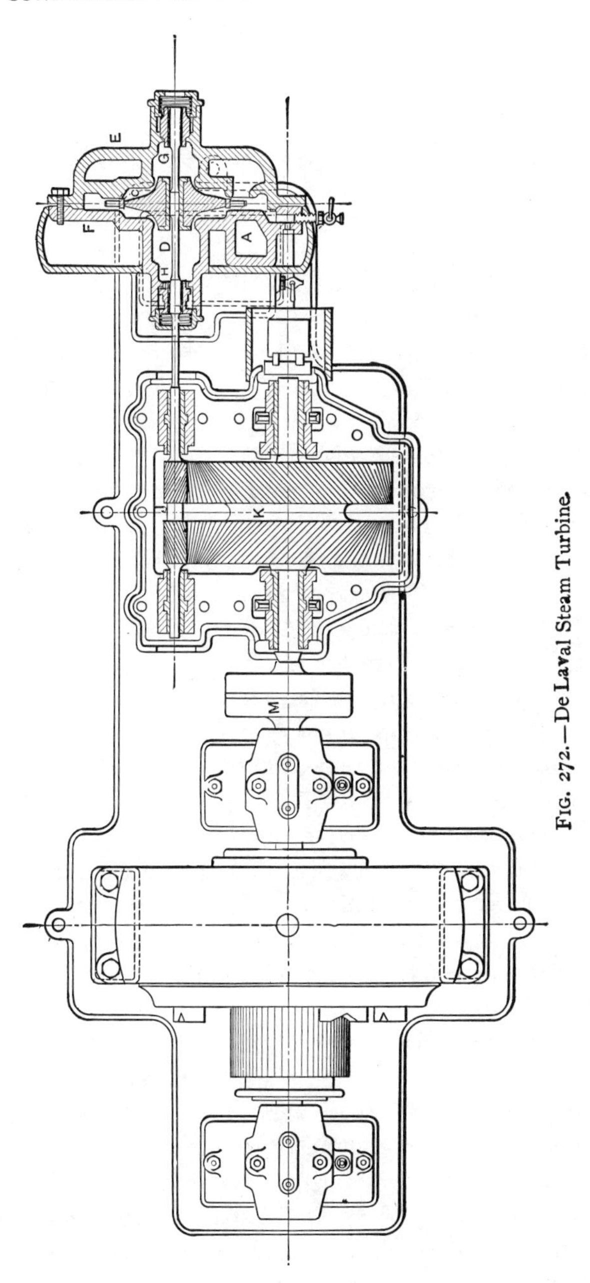

FIG. 272.—De Laval Steam Turbine.

C, which is mounted upon a flexible shaft D. Nozzles are arranged around the wheel and may be thrown in or out of

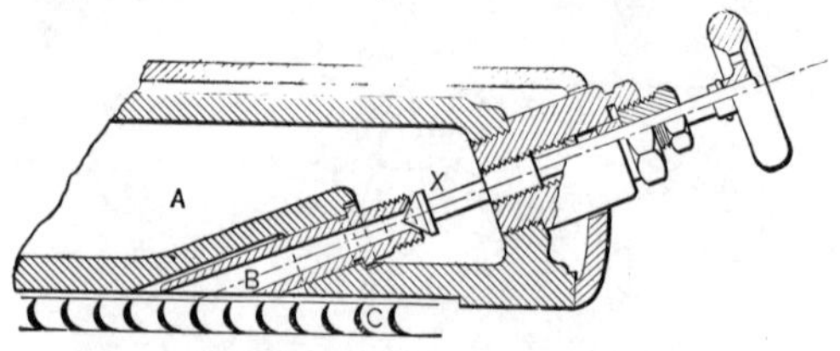

FIG. 273.—De Laval Steam Turbine Nozzle.

service by valves X. The high velocity of the steam leaving the nozzle gives a high rotative speed to the wheel, varying from 10,000 to 30,000 revolutions per minute, the higher speed being for the smaller wheels.

These speeds require that the hub of the wheel be made of the best forged nickel steel. Into milled slots in the periphery of this hub are placed the bulb ends of drop-forged wheel vanes or buckets. The wheel cases E and F fit down closely to the hub to check the speed of the wheel should accident occur. The shaft is flexible to allow the wheel to turn about its center of gravity and in its proper plane. The tendency for a body to revolve in this way is very great at high speeds, and to prevent vibrations and rupture this small shaft has been found necessary. The small turning moment at this speed renders this possible. To accommodate any bending in the shaft the bearing G is made with a spherical end, while H is free to move in any direction. The springs in the casings of these two bearings prevent steam from leaking out when running non-condensing, or air from entering when running condensing. The high speed of the turbine shaft is reduced from 10 to 1 by two pairs of spiral-gears JK. These gears are made with the teeth making an angle of 90 degrees with each other and prevent end play. The coupling M unites the shaft of the gear wheel to the dynamo shaft.

A test of a De Laval turbine giving 266.7 B.H.P. shows a steam consumption of 17.35 lbs. per B.H.P. hour, with 155 lbs. of steam and 25.5 inches of vacuum. The revolutions of the turbine were 9000 per minute.

INDEX.